I0468111

TEACHING MATHEMATICS

Favorite Topics and Techniques

by

William Bradley Martin

Copyright 2013 © William Bradley Martin
All rights reserved.
ISBN-13: 978-1482568417
ISBN-10: 1482568411

Table of Contents

Introduction

On the first day of class I tell the students that I have

bachelor's degree in mathematics and a Masters in applied

mathematics. The applied mathematics will be used to help

you see how mathematics can be used. But equally, I like so

called pure mathematics, such as proving the square root of 2

is irrational. This book has both themes and is based on

teaching for 30 years at a community college in the

southwest—everything from arithmetic through differential

equations.

Thirty years of experience with students has given

me some strong opinions on pedagogy and what mathematics

to emphasize. I have seen the ability of students to work with

fractions steadily decline for at least twenty years. Many high

school graduates with algebraic credits have no knowledge of,

and see no need to organize algebraic processes into vertical steps. I discuss pedagogy and some nice enrichment mathematics at a fairly low level in Part One, "Elementary and Intermediate Algebra."

The teaching of trigonometry and calculus has stimulated my own research interests which are number theory and mathematics applied to physical problems. Much of this, at an undergraduate level, is in Part Two, "College Algebra, Trigonometry, and Calculus."

I have tried to find the most important and simple material that is still interesting. I hope that teachers and those wishing to learn will find the two themes, favorite topics and teaching techniques useful.

William Bradley Martin, Tucson, 2013

Part I: Elementary and Intermediate Algebra

1. An Astounding Encounter

I faced 18 students, half traditional college age and half above that, ranging up to about age 40. It was a pre-algebra class, which is mostly arithmetic with a dash of algebra and some geometry formulas.

"What is the formula for the circumference of a circle?" I asked. Several of the older students responded quickly, "Pi times D." I was gratified but wanted to make sure that all understood.

"That means that it's about three times as far around a circle as it is across it." Now began my most memorable encounter with students in my whole career. There was absolute silence -- I began to sense doubt. Questioning further, I found (believe me!) firm disbelief from everyone! Even from the three or four who had parroted the, "Pi times

D" response. How could this be? What was going on in their heads?

I did not know how to go on. One can't proceed to typical problems if they don't even believe (or understand) the relationship of diameter and circumference. What is the point? With these students, a proof based on some limiting process was way beyond reach.

I attacked the disbelief with a diagram -- a large picture of a square circumscribing a circle. I pointed to a diameter and showed that it equaled one side of the square. Triumphantly I said, "Going around the square (while moving the chalk) you must travel four times as far as the diameter. But going around the circle, and cutting off the corners must be shorter than this." I joked that I have cut across many lawns in my youth and know this from experience!

Continuing, "So, the circumference has to be less the four times the diameter. In fact, it's Pi times the diameter and you know that Pi is about 3.14." I could see this made a real impression! Another astounding fact: I bet they had never seen this!

My universally doubted statement did not seem crazy. Of course, there could have been some initial confusion about three versus the irrational Pi. But I emphasized the phrase "about three." No, they really doubted the relationship, the whole idea. They had no picture of the relationship between the diameter and the circumference. They had been very resistant to the whole thing and, as I said, they seem to have never seen a square around a circle.

After class and many times since, I have wondered about this glimpse into what we think students understand. Knowing formulas and even being able to solve typical problems based on a formula does not mean that students really understand a relationship. You can give a diameter or a circumference, and the student is trained to somehow come up with the missing quantity. But do they understand? I have given the following problem to much higher level students, a trigonometry class:

A bicycle wheel 26 inches in diameter rolled along the pavement through 5 revolutions. How far did the center of the wheel travel?

Many (over half!) -- stumble badly and can't do it!

What is the moral here? I think it is this: Don't just teach formulas and give boring problems. Give problems that probe real understanding. Above all, ask questions that force understanding and thinking.

2. The Language of Fractions

I've seen the ability of community college students to manipulate fractions go steadily down in thirty years of teaching. There is a visceral repulsion to finding common denominators or even simplifying fractions. Students have been taught not to cancel!

One contributing cause is too great an addiction to decimals, which is partly due to calculators. I counter the aversion to fractions with examples like this:

At a picnic there are 7 pies and 13 families. To share equally each family would get 7/13 of a pie. How do we know this is right? I write:

$$13 \text{ family} * \frac{7}{13} \frac{pies}{family} = 7 \text{ pies}$$

(Cancel the word family on top and bottom, leaving only the word pies)

Thus all the pies are used up and of course each family gets the same amount, 7/13 of a pie.

Can you deal with these quantities as easily using decimals? Well, 7/13 equals:

$$.\overline{538461}$$

This a 6 digit repeating decimal! Can you do the above check as easily? If you enter the following product on a calculator:

$$.538461 \times 13$$

you get 6.99993, not 7! Of course, if you multiply the calculator's decimal version of 7/13 and multiply the calculator will round correctly to 7. Will the same fortuitous rounding occur with all fractions?

It's very tiresome to write $.\overline{538461}$, and almost impossible to get a word processor to type a line over the digits. The fraction 7/13 is simple and beautiful by comparison!

Also the student, handicapped by fraction illiteracy, loses not only facility with handling quantities but also understanding. For example if one more family came to the party, making 14, then in the same way we get $7/14 = 1/2$ pies per family. The amount each family gets must be slightly less than before. Also look at the reciprocal situation: 14 pies and 7 families implies $14/7 = 2$ pies per family. We see the power and convenience of fractions. It all makes sense, but using decimals requires tedious conversions by hand or a calculator and the students loses understanding. The student loses the meaning of fractions and the relationships between them.

Why have we allowed so many students to become addicted to decimals? Most fractions have ugly decimal versions. Of the 25 reciprocals of integers:

½, 1/3,2/3,…1/26

17, almost 70% have decimal patterns that go on forever! Many of these fractions have very long repeating patterns. The exact decimal version of 1/17 is

$.0\overline{5882352941176}47$, a 16 digit repeating pattern! , 1/23 has a 22 digit repeating pattern! These two examples give a hint of a rule about the length of the pattern. This is a great area for our students to explore and is discussed in the next section.

The decimal patterns of so many fractions are so long that you will never see the pattern on a calculator. Thus you will never see the correct value on a calculator of many of these basic reciprocals. Consider just the family of fractions, n/23. The only decimal way to write exactly any of the 22 fractions:

1/23, 2/23, 3/23,…22/23

is to write a decimal with 22 digits. Each fraction must have a line over all the digits!

The ancient Egyptians, creators of much geometry and the still existing pyramids, never had a good symbolism for fractions. They had good symbols for reciprocals, but with the single exception of $2/3$, their expressions for other fractions was very clumsy. For $3/4$ the Egyptians wrote the equivalent of $1/2+1/4$. For $5/7$ they wrote $1/2+1/5+1/70$! Students should appreciate the convenience and utility of fractions.

The next section shows that there are fascinating and deep aspects to fractions. There are many aspects not discussed in this book: Farey Series, Continued Fractions, Approximating irrational numbers, and Simpson's Paradox . Countering students' phobia about fractions, I say in class , "Let us love and embrace fractions."

3. The Decimal Version of Fractions

To see the decimal version of a fraction roll out of a calculator has always been fascinating to me. Some decimals are short and some are very long and repeat forever. For example:

$9/16 = .5625$, has only 4 digits and is terminating.

But

$$9/17 = .\overline{5294117647058823}$$

is infinite, a repeating pattern of 16 digits!

In fact, since calculators only display 12 or fewer digits, it is impossible to recognize the 16 digit pattern—I did the decimal expansion on a computer using Mathematica.

In this section, I will look at some facts about the decimal expansion of fractions. The emphasis will be on the lengths of the repeating patterns in non-terminating fractions. Terminating fractions are those with denominators that factor into powers of 2 and/or 5 only. Denominators with prime factors in addition to 2 and 5 have infinite repeating patterns. We will state some rules about these infinite patterns. Proofs will not be given but some justification will. Later we will apply these rules to many fractions.

Look at these two fractions:

1. $\dfrac{51}{99} = .\overline{51}$ 　　　　　 2. $\dfrac{264}{999} = .\overline{264}$

These fractions illustrate two facts:

Fact 1: Proper fractions with nine's in the denominator repeat forever with the same digits as in the numerator.

The second fact has to do with the length of the repeating decimal in all fractions, not just those with all

9's in the denominator. Let us reduce our previous two examples:

Example 1. $\dfrac{51}{99} = \dfrac{17}{33} = \dfrac{17}{3 \times 11}$

Example 2. $\dfrac{264}{999} = \dfrac{88}{333} = \dfrac{88}{3^2 \times 37}$

I have factored the denominators into prime factors. These prime factors dictate the length of the repeating pattern. Fact 2 is about the length of the repeating pattern and applies to all fractions, including improper ones:

Fact 2: For any fraction k/d, the length of the repeating pattern is n where n is the lowest power in $(10^n - 1)$ that is divisible by all the prime factors (excluding 2 and 5) of d. Note that $(10^n - 1)$ is a string of n 9's. So Fact 2 has a connection to Fact 1.

Let's see how this fact applies to our examples. In example 1, the reduced denominator, 33 factors into 3 times 11. Both these numbers divide into $(10^2 - 1)$, and two is the lowest n in the expression $(10^n - 1)$ for which this is true. Therefore 17/33 has a two digit repeating pattern.

In example 2, both 3^2 and 37 divides into $(10^3 - 1)$ but no lower power of 10. So this fraction has a three digit pattern. Note that $(10^3 - 1) = 3^3 \times 37$. This may help you predict the next fact which concerns denominators that are powers of primes.

Look what happens to the decimal expansion when we increase the power of three in these denominators:

$$\frac{26}{3} = 8.\overline{6} \qquad \frac{26}{3^2} = 2.\overline{8} \qquad \text{but} \qquad \frac{26}{3^3} = .\overline{962}$$

Why did the third power of 3 suddenly produce a three digit pattern? Because 3^3 divides evenly into $(10^3 - 1) = 999$ but no lower power of 10 in $(10^n - 1)$

Let's expand fact 2 to denominators that have powers of primes.

Fact 3: For any fraction $\dfrac{k}{p^e}$, where p is a prime other than 2 and 5, the length of the repeating pattern will be n where n is the lowest power in $(10^n - 1)$ (that is divisible by p^e.

This fact can be extended to products of powers in the denominator: We must find the lowest n in $10^n - 1$ such that all prime factors in the denominator with their exponents divide into. The repeating pattern will then be of length n. Some problems with answers follow.

Student Problems, (Answers Follow)

1) Consider fractions with denominators consisting of powers of two and/or five along with other primes. In particular, how long is the repeating pattern and is there anything different about the patterns compared to fractions with no 2's or 5's in the denominator? For example,

$$\frac{2}{7} \text{ versus } \frac{2}{5*7}$$

2) There is something very simple we can say about the maximum length of the repeating pattern in any fraction k/d. In terms of d what is the maximum length possible? Why?

3) Explain the length of the repeating pattern in the

two cases below.

Case 1: $\frac{2}{297} = .\overline{006734}$, a six digit repeating pattern.

Case 2: $\dfrac{56}{41} = 1.\overline{36585}$, a five digit

repeating pattern

Answers

1. The repeating pattern is still dictated by Fact 3. The only difference is that the repeating pattern does not start immediately after the decimal point. For example:

$$\frac{172}{2^2 * 5 * 3} = 2.8\overline{6} \text{ , a one digit repeater}$$

because of the 3.

$$\frac{172}{2^2 * 5 * 3 * 11} = .26\overline{06}, \text{ a two digit repeater}$$

because of the 11.

2. The repeating pattern will be at most (d-1) digits long. In the division process of converting to a decimal, only so many remainders are possible. In fact, only the remainders from the set: {1,2,3,...,d-1) are possible. Once a previously

encountered remainder occurs, the process will start repeating

and continue with the same decimal pattern forever.

3. *Case* 1: $297 = 3^3 * 11$ and the smallest n for which 3^3 and 11 both divide $(10^n - 1)$ is n =6.

Case 2: 41 divides into $(10^5 - 1)$.

One final point. There is an important theorem from

number theory that says if p is a prime and a is relatively

prime to p then

$$a^{p-1} - 1 \text{ is divisible by p.}$$

This is called Fermat's little theorem. This implies that for all

primes not equal to 2 or 5:

$$10^{p-1} - 1 \text{ will be divisible by p.}$$

The implication for decimal patterns is that all

fractions with prime denominators p (not equal to 2 or 5) will have repeating patterns of at most (p-1). This was also guaranteed in another way by problem 2 above.

4. Mediants

If $\dfrac{a}{b}, \dfrac{c}{d}$ are two fractions then the mediant M of

them is given by

$$M = \frac{a+b}{c+d} \ .$$

In other words add the numerators and the denominators. This erroneous way to add two fractions has, nevertheless, many amazing and useful properties. The contrast with the correct sum,

$$S = \frac{a}{b} + \frac{c}{d} = \frac{ad+bc}{bd}$$

provides good practice for our students. The comparison of the mediant with the average will be another fruitful comparison.

In this section, we will always assume that:

1. $\dfrac{a}{b} < \dfrac{c}{d}$. This implies that ad< bc.

2. That all integers a, b, c, and d are positive.

As its name implies, the mediant is between the two fractions, $\dfrac{a}{b}$ and $\dfrac{c}{d}$. The proof of this between--ness property and a good exercise for students starts by looking at the difference:

$$M - \frac{a}{b} = \frac{a+b}{c+d} - \frac{a}{b} = \frac{bc - ad}{b(c+d)}$$

But since bc > ad, this is a positive fraction. So, M is greater than $\dfrac{a}{b}$.

The second half of the proof is to show that $\dfrac{c}{d} - M$ is positive and is left to the reader. It's time for some examples.

1.Find the average and the mediant of the following three pairs of fractions. Verify the between--ness property of the mediant and state whether the average or the mediant is greater. Note: Do not reduce $9/12$ in part c.

$$a) \frac{2}{3} \text{ and } \frac{3}{4} \quad b)\frac{1}{5} \text{ and } \frac{3}{4} \quad c) \frac{2}{3} \text{ and } \frac{9}{12}$$

Assuming you worked the problem, did you notice that the mediant in problem 1a was not the same as in 1c even though the original fractions have the same value. The mediant is unique only for reduced fractions.

2. It is shown below that the average of $\frac{a}{b}$ and $\frac{c}{d}$ is greater than the mediant if the first (lesser) fraction's denominator is greater. Conversely, the mediant is greater than the average when the greater fraction's denominator is bigger. Which condition do the above cases satisfy?

The Average Compared to the Mediant

Again, we assume $\dfrac{a}{b} < \dfrac{c}{d}$ and all letters are positive integers. Look at the difference

$$Average - Mediant = \frac{ad + bc}{2bd} - \frac{a+b}{c+d}$$

Combining these fractions, one can show that numerator is equal to (b-d)(bc-ad) . The second factor of the numerator is positive by assumption. So, the average is bigger only if b > d, i. e. only if the smaller fraction's denominator is greater than the larger fraction's denominator. Conversely, the mediant is greater if the second (greater) fraction has a larger denominator. This rule applies to the forms of the fractions, whether reduced and unreduced versions. When will the average equal the mediant?

The Determinate Property of Mediants

With the same fractions $\dfrac{a}{b}$ and $\dfrac{c}{d}$ and again with

$\dfrac{a}{b} < \dfrac{c}{d}$, form what I call, "The Determinate of the

Fractions":

$$\begin{vmatrix} a & c \\ b & d \end{vmatrix} = ad - bc$$

Note that the order of the fractions makes a difference. When I say the determinant of the fractions f_1 and f_2, the first column of the determinant should be the numerator and denominator of f_1. If ad-bc equals -1 then the mediant of the fractions $\dfrac{a}{b}$ and $\dfrac{c}{d}$ has the wonderful property that it is the unique rational number between $\dfrac{a}{b}$ and $\dfrac{c}{d}$ with the smallest denominator. A proof of this property can be found in a "Wikipedia" article under

"Mediant." From now on, we assume all fractions are reduced. Look again at the problem 1a above. The determinate of the fractions is − 1. The mediant is 5/7. No fraction with denominator less than 7 is between the original fractions. I continue now with some student problems exploiting the above ideas.

<u>Student Problems Continued—No Answers</u>

3. In problem 1a, the mediant of $\dfrac{2}{3}$ and $\dfrac{3}{4}$ was $\dfrac{5}{7}$.

The determinant of the original fractions was −1. What are the determinants of the two sequential pairs:

$$\dfrac{2}{3} \text{ and } \dfrac{5}{7}? \text{ , and of } \dfrac{5}{7} \text{ and } \dfrac{3}{4}?$$

4. Suppose the determinant for two reduced fractions $\dfrac{a}{b}$ and $\dfrac{c}{d}$ is −1. Let the mediant of them be as follows:

$$\dfrac{a+b}{c+d} = \dfrac{f}{g}$$

So, we now have the three fractions in the order:

$$\frac{a}{b}, \frac{f}{g}, \frac{c}{d}$$

Show that the two possible sequential fraction's determinants are such that:

$$\begin{vmatrix} a & f \\ b & g \end{vmatrix} = -1 \quad \text{and} \quad \begin{vmatrix} f & c \\ g & d \end{vmatrix} = -1.$$

In other words, the -1 value of the determinant of the fractions always perpetuates when you form the mediant.

5. Let's exploit the property of mediants in problem 4 to get a mediant series.

a) Let's start with two new fractions, 1/2 and 2/3. Use the mediant to find the rational number between 1/2 and 2/3 with the smallest denominator.

b) Now write the three fractions together in one line:

$$\frac{1}{2}, \ (\textit{mediant from part a}), \ \frac{2}{3}$$

Show that the determinant of ½ and the mediant (in that order!) is −1 . Show that the determinant of the mediant and 2/3 is also −1.

Chains of Mediants

If we continue with these fractions and keep taking the mediants of every sequential pair we get the chains of sequences:

$$\frac{1}{2}, \frac{2}{3}$$

$$\frac{1}{2}, \frac{3}{5}, \frac{2}{3}$$

$$\frac{1}{2}, \frac{4}{7}, \frac{3}{5}, \frac{5}{8}, \frac{2}{3}$$

$$\frac{1}{2}, \frac{5}{9}, \frac{4}{7}, \frac{7}{12}, \frac{3}{5}, \frac{8}{13}, \frac{5}{8}, \frac{7}{11}, \frac{2}{3}$$

Note that the determinant of every sequential pair of fractions, taken in increasing order, is −1. Each fraction between any two fractions is the unique fraction with the smallest denominator between the two. The number of fractions is increasing rapidly. If S_n is the number of fractions in the nth step with $S_1 = 2$, then

$$S_{n+1} = S_n + (n-1)$$

And

$$S_n = 2^{n-1} + 1 \text{ , for } n \geq 1$$

Think of it. One can make ordered, infinite chains of fractions starting with any two fractions for which

ad-bc $= -1$.

Every chain of fractions is such that every sequential pair has a determinant equal to minus one. We are not limited to proper fractions. For example, start with the pair $\frac{10}{7}$ and $\frac{13}{9}$. The determinant of the fractions is -1. Form chains of mediants. The tenth chain will be a sequence of 513 ordered fractions. For every possible sequential triple of these

513 fractions, the middle fraction is the unique rational number between the two outer fractions that has the smallest denominator.

5. Games with Percents

For my Algebra One and lower level students, I love to give this problem:

Sue makes $40,000 per year. One day, the boss says that times are tough and tells her she has to take a 10% cut. A year later, the economy is better and the boss says, "Sue, I'm giving you a 10% raise, congratulations!" Is Sue back to her original salary? If not, what is the difference?

This problem shows the trickiness of relative change. Letting r be the fractional change in her salary which is the same at both steps. The result of the two steps can be generally calculated as

$$40000 \, (1-r)(1+r) = 40000(1-r^2)$$

We see that she must make less after two steps. In this case by the factor $r^2 = .1^2$. Multiplying with 40000 equals $400. Sue's salary is less by $400.

Also the order of the salary steps makes no difference. Sue could have received 10% increase first and then a 10% cut—same result $400 less. It seems unfair to our employee but at least mathematics, in particular the commutative principle, helps us understand why!

Here is another way to look at it for students who've had infinite geometric series.

After a salary cut of r, her salary is reduced by the factor (1-r). Then, to increase back to the original salary, Sue would need to increase by the reciprocal of (1-r). But:

$$\frac{1}{1-r} = 1 + r + r^2 + r^3 + \ldots$$

Thus, for example, if Sue originally took a 10% cut, then a never ending series of raises which are increasing powers of 10% would get her back to her original salary! Of course, a onetime increase of $.\overline{1} = 11.\overline{1}\% = 1/9$ would also work but that is not as much fun.

6. Five Minutes of Logic

It's amazing how valuable a little logic is. Many theorems have the structure:

If P, then Q,

where P and Q are statements. There are three possible alterations of this structure: The Converse, The Inverse, and The Contrapositive. We can give students the structure of all four in a table:

THEOREM	SYMBOLICALLY
Original Implication:	If P then Q
Converse of Theorem:	If Q then P
Inverse of Theorem	If not P then not Q
Contrapositive of Theorem:	If not Q then not P

In class I use theorems that are true. It's fun to start the lesson with non-mathematical theorems :

1) If you live in Tucson, then you live in Arizona.

2) If you are a murderer, then you killed someone.

Students can easily see that that the converse of (1) is false, but surprisingly many think the converse of (2) is always true. Many have to be reminded what murder is. I use the word "murderer" to denote an actual murderer and not merely one convicted. Students who think the converse of (2) is true have trouble thinking of counter--examples, and different scenarios (self-defense, negligence). It is surprising how long it takes and how reluctant students are to accept these examples. Be sure to go through the inverse and the contrapositive of both theorems. We are setting the students up to see the pattern in the equivalences of the four possibilities. Many don't see the importance of definitions, and their importance for discussion. It's easy to see why people disagree about things, if they don't see the importance of definitions and basic logic.

Here are some math theorems that require almost no background:

3) If a number is a multiple of 4, then it is even.

4) If ABC is a right triangle, then the sum of the squares of two sides equals the square of the remaining side.

Many students see easily that the converse of (3) is false. They are not sure about the converse of (4), the only case of the four theorems we've had where the converse is true.

Venn diagrams give very convincing proofs of all the above. A valid theorem, "If P then Q", is shown with a circle for P inside a rectangle for Q. It is easily seen that the converse, "If Q then P", is in general false—If you are in the rectangle Q you are not necessarily inside the circle P. The same applies to showing the general falsity of the inverse. The diagrams also are very convincing for believing that the contrapositive is always true when the original theorem is. We want students to appreciate that a favorite technique to prove a theorem is to prove the contrapositive—and just as good as proving the original theorem.

Good examples, mathematical and otherwise, should convince students of the validity and usefulness of basic logic. Armed with just a little logic, they can see the dangers of assuming converses and inverses in math, and in every subject.

7. Our Number System

You can't follow a football game well if you don't know the players and what they do. In math the players are numbers and there are many types and they do different things, just as on a football team. For example only rational numbers are solutions to linear equations. There are many theorems on the zeros of polynomials that depend on the different classifications of numbers: integers, rational or complex numbers. I do a simple 15 minute lecture on our number system at some point in all my classes. Here it is.

Let's go way back. The oldest pyramids are 5000 years old, but 35,000 year old bones exist on which cavemen carved tally marks. On one such a wolf bone, found in the Czech Republic, there are 37 marks in groups of 5—4 vertical strokes and a cross stroke, tally marks. Somebody was counting to thirty-seven 35,000 years ago! Gradually these line marks evolved into symbols for individual numbers. Here is a dramatic demonstration you can do in class that shows how our modern symbols may have come about.

Our caveman ancestors drew one vertical line to indicate one thing—you do it on the board. For two things (goats, whatever) they drew two horizontal lines—you do likewise next to the vertical line, Similarly say three horizontal strokes for three and draw them. You should have:

Then, with a flourish, draw curved lines connecting the caveman's two lines and do the same for the three marks--making them look the modern symbols for 2 and 3 respectively. Say at the same time, "This evolved into our one, two and three!"

Did our 1, 2 and 3 really come about this way? Who knows? I once had a student from the Far East who said the history was the same for them--unexpected corroboration right it class!

We draw a box around these numbers, extended infinitely and labeled as so:

```
┌─────────────────────────────────────┐
│          Counting Numbers           │
│                                     │
│            1,2,3,4,…                 │
│                                     │
└─────────────────────────────────────┘
```

But what about negatives and the very important number, zero? Why do we need zero anyway, a symbol for nothing? Let's talk about zero--it provides a pleasant diversion. I contrast the Roman Numerals with our system of numbers with a table, next page. Make two columns, with headings, "Our Numbers" and "Roman Numerals". Put combinations of "6's and 0's" in the "Our Numbers" column, as shown. We are showing the power of place value also. Ask the students to state the Roman equivalents. Eventually the filled in table looks like this:

Our Numbers (Don't give its name yet !)	Roman Numbers[1]
6	VI
60	LX
66	LXVI
606	DCVI
660	DCLX
666	DCLXVI
6066	MMMMMMLXVI

This class activity shows the tremendously greater facility of "Our Numbers"—still unnamed. Go further: What if we had to multiply in Roman numerals DCLXVI by DCVI? Or even worse, DCLXVI ÷ DCVI ! Yet, the Romans must have done it.

The combination of place value and a symbol for nothing were very powerful inventions. The Romans, Greeks, Egyptians and many other civilizations never had this idea, and consequently, could not do basic arithmetic nearly as easily as we can. I think students should know this.

---

[1] The evolution of the Roman System was complex. Different symbols for the same quantities were used. The six M's in the last example may never have been used.

Who invented zero? Students frequently say the Greeks or native people of the New World. But Western Civilization is indebted to the Hindus for zero. Continuing the usage of zero and its transmittance to Europe is attributed to Arabs who traded with the Hindus. So our number system is properly called the "Hindu-Arabic System."

Continuing with the different types of numbers, we can motivate the acceptance of negative numbers (and the expansion of Our Number System) with an apocryphal story of the great Diophantus. This ancient mathematician solved very tricky equations, whose solutions are limited to integers, and which are now called Diophantine problems. It is said that Diophantus once considered the very simple problem of solving (in our notation) the equation:

$$x + 3 = 2$$

Diophantus struggled to find x. He stated, "The solution does not exist or makes no sense." One wonders from the second half of his statement how close he came to

creating negative numbers. How would our history have been

different if negative numbers had been accepted and used,

1500 years before they actually were? So, adding the negative

whole numbers and zero, we have the set of Integers. Our

diagram on the board expands to:

Integers

...,-3,-2-1,0

Counting

Numbers

1,2,3,4,...

The expansion of our Number System continues. When a person died, his property had to be divided. Thus, Rational Numbers were born very early—thousands of years before zero and negative numbers. We define Rational Numbers, as numbers that can be written the ratio of two integers. Our diagram grows to:

RATIONALS

Definition: Can be written as the

ratio of two integers

Integers

...,-3,-2-1,0

Counting

Numbers

1,2,3,4,...

53/7, -2358/401,

.432,.666666.....

Challenge the students. "Is .432 rational? How would you prove it?" Confronted with $.\overline{6} = .66666...$, many think the number is not rational. They have seen infinite decimals many times on calculators but the connection of the word "Rational" to the ratio of two integers is usually non-existent.

Next we have a big jump, the Irrationals (see also the next section, Number Types and Incommensurables.)

The first irrational recognized was $\sqrt{2}$. How can we resist telling students that it was proven irrational by the Greeks and giving the argument using even and odd integers? It's totally accessible to our students. While you'll never see $\sqrt{2}$ on a grocery bill (this usually gets a smile), the simplest square with unit sides has $\sqrt{2}$ as its diagonal. Our board now has a box to the side of the previous one labeled, "Irrationals" with $\sqrt{2}$, and other examples such as $\sqrt[3]{7}$, sums of Rationals and Irrationals, and π written in it. We put a box around everything and write, "Real Numbers". It's nice to have one word for everything! Emphasize that the reals are all the numbers on a number line. Draw a number line with $\sqrt{2}$ and π correctly located.

Your students may hope that they won't need any more numbers, but of course they do. We can motivate the next big group with a simple looking problem. (In analogy with the expansion to negative numbers before):

Solve:

$$x^2 + 1 = 0$$

We are forced to accept *i and* $-i$. Also simple quadratic equations like

$$x^2 + 6x + 10 = 0$$

lead to the solutions -3-i and -3+i. The final diagram with labels and arbitrary examples follows on the next page.

COMPLEX NUMBERS

$$3+i \;,\; 12+7.8i \;,\; 4\pi - \frac{22}{3}i$$

Real Numbers		Imaginaries
		i, 5.8i

Rationals   **Irrationals**

$$\frac{53}{7}, .\overline{321} \quad \frac{\sqrt{2}}{3}$$

$$\sqrt[3]{7}+5$$

Integers

$$\pi$$

$$...-3,-2,-1,0 ...$$

Counting

Numbers

$$1,2,3,4,...$$

-2358/401,

.432,.666666

'8. Number Types, Incommensurability

We hope that our students learn and appreciate the following.

A number is rational if and only if its decimal version terminates or repeats.   (1)

To appreciate this statement, students must be grounded in the logic that converses are not always true. A previous section, "Five Minutes of logic", should help. Let's look at the proofs of this double implication.

I. If a number is rational, then its decimal version terminates or repeats.

PROOF: A rational number a/b, terminates if b factors as products of two and/or five only. Suppose b has factors other than 2 or 5. Dividing by the integer b means that there are only (b-1) possible remainders. This guarantees that a remainder has to reoccur by the (b-1)st iteration in the division process. When a previously seen remainder reoccurs, the process will then cycle through those remainders and the decimal number will repeat forever in the same pattern. Thus, if a number is rational, its decimal version terminates or repeats. In fact, this argument suggests the additional fact:

All repeating rationals a/b have a of period of (b-1) or a smaller period that divides into (b-1).

The other half of the double implication follows.

II. If a number has a terminating or repeating decimal, then it is rational.

PROOF: If x is terminating, then obviously it is rational.

Suppose x repeats forever. Let's look at a specific example, which will suggest the general argument. Suppose x starts repeating after 2 digits with a 3 digit pattern.

$$x = .57\overline{234}$$

but $10^{2+3}x = 57234.\overline{234}$

While $10^2 x = 57.\overline{234}$

So $10^5 x - 10^2 x = 57177.$

$$x = \frac{57177}{10^5 - 10^2}$$

We have proved that this x is rational. The process easily generalizes:

Let x, be the repeating decimal number: x = . $D\,\overline{R}$

where: D is the string of digits with length d that occur before the decimal starts repeating, and $\overline{R}$ is the repeating string of digits of length r that repeats forever.

Then $10^{d+r}(x) = DR.\overline{R}$

and $\qquad 10^d x = D.\overline{R}$

Subtracting the bottom equation from the top implies that

$$x = \frac{DR - D}{10^d(10^r - 1)}$$

This proves that x is the ratio of two integers and is therefore rational. The case of an integer with a repeating decimal is obviously rational also. We have proven the second half of the double implication.

Statement (1) also implies the following:

A number is irrational if and only if its decimal pattern goes on forever and never repeats.

This idea should help students understand that special symbols such as $\sqrt{2}$ and π are required for these irrational numbers.

Irrationals are Incommensurable with Rationals

How can we impress students with incommensurability? I give the standard proof that $\sqrt{2}$ is cannot be the ratio of two integers. (Assume $\sqrt{2}$ = a/b, a fraction reduced to lowest terms and get a contradiction).The students are not moved! There must be a dramatic way.

Start by drawing and labeling with their lengths two commensurable line segments A and B whose ratio is close to the square root of 2:

A _____ 1.4

B _____ 1.0

We can say that A is 14 (1/10ths) and B is 10 (1/10 ths). I can measure the lengths with a ruler letting A be 1.4 meters and B one meter To be dramatic, cut the first into 14 equal pieces and the second into 10. We say these lengths have a common measure, namely 1/10 of a meter or a decimeter.

Do the same thing (drawing line segments as before) with A = 1.41meters and B= 1.0 meters. In this case, the common measure is 1/100 th of a meter (a centimeter). Make an outline sketch of cutting the first into 141 pieces and the second into 100 pieces.

"But look, try to do the same thing with $\sqrt{2}$ and 1.0." (Draw the picture):

$A = \sqrt{2}$: _____

$B = 1$: _____

Suppose we assume these lengths do have a common measure that is very small. Call the common measure "zinches". Why not? You can make several "zinch" cuts (very small!) on both line segments.

So let $\sqrt{2}$ be a zinches long. Suppose 1.0 equals b zinches, where a and b are integers. But then:

$$\frac{Segment\ A}{Segment\ B} = \frac{a\ \text{zinches}}{b\ \text{zinches}} = \frac{\sqrt{2}}{1}$$

or   (Canceling the zinches)

$$\frac{a}{b} = \sqrt{2},$$

But, of course this contradicts our ancient proof that $\sqrt{2}$ cannot equal the ratio of two integers. You are 2000 years out of date if you think $\sqrt{2}$ can! The ancient Greeks were uncomfortable with incommensurables. Our students should feel this too.

9. *Integers Only*

Here is a problem which is unlike any you'll see in an elementary algebra textbook.

Find some integer values for x and y that satisfy the following equation.

$$3x - 2y = 1$$

Number theory says that since 2 and 3 are relatively prime, there will be infinitely many pairs of integer solutions. But the student needs to attack this problem by simply trying integers--guessing. Without much trouble, the guess x=1, and y =1 may be the first solution found. But there is a layer of beauty below this, an interesting pattern, which we want our students to find. The students need some guidance. Give the hint: There are more than one pair of integers (x,y) that work—find many pairs! Then require students to: Put their results in a table in order of **increasing x.** Can they find a

pattern? Make the students state it!

So how would an untrained mind react to the whole

problem? The restriction to integer solutions is off putting for

some. Can we get our students to boldly start guessing

integers?—That is what we want! In addition to (1,1), they

hopefully could find other solutions. If a student tries x = 2,

it is a dead end--no integer y works. Will she be bold enough,

persistent enough to try x =3? The reward is the easily found

y =4 So (3,4) is another pair. How likely after this could

he/she find the next one? Students need to be pushed!

Another pair is (5,7). The vast majority of students would be

satisfied—We should not let them feel this way, yet!

It's very important to force the students into

organizing their work. Only in this way can they see a pattern,

the goal of mathematics. We included the important proviso

that the table be in order of increasing x. Here is the result so

far:

X	Y
1	1
3	4
5	7

Now the climax. What is the pattern? Tell the students to look! Hopefully, they write:

"x is increasing by 2 while y is increasing 3".

Conclusions should be clearly written and succinct! Maybe the student also realizes that the increments on x and y are the opposite of the coefficients on x and y. There is always another layer of beauty to a good problem.

There is so much more to this subject, of course. Look at the ratios of y to x as both increase—what will the ratio go to? Look at negative integers solutions. Graph the solutions.

The equation

4x -2y =1

has no solutions in integers. Can the students find the

criterion for when the equation

$$ax - by = 1$$

does have solutions?

The restriction to integer values, Diophantine

problems, opens a beautiful subject, which at this level is well

within students' reach, but is rarely touched in elementary

mathematics courses. Is it our fault if students think

mathematics is a set of rules, to be unthinkingly applied to

always similar exercises? And discovery playing no part? Yes,

if we don't show them good problems with layers of patterns

like this problem.

10. Show the Answer Works

When I solve simple linear equations in class like:

$$2x + 1 = 6$$

I also very carefully show the students how to write

out a, <u>Show The Answer Works</u>, part. In this case the answer

is 5/2. Here is the <u>Show The Answer Works</u>, part which is

just as important!

<u>SHOW THE ANSWER WORKS</u>

$2x+1=6$

$$2\left(\frac{5}{2}\right)+1\overset{?}{=}6$$

$$5+1\overset{?}{=}6$$

$$6=6!$$

I don't make students do this on every problem, just

occasionally, and especially at the beginning of the course.

Years of experience with students have forced me to be very

specific about how they write out this demonstration. I am very strict that students must start by writing the original equation, followed in the next line by substituting the supposed solution. In this line, and in subsequent lines, I also require a question mark over the equals sign until an identity is reached. Also very important is that each side should be simplified independently. Here is a list of requirements which must be followed to the letter and some of the ways the requirements are violated by students.

SHOW YOUR ANSWER WORKS-REQUIREMENTS

AND COMMON STUDENT VIOLATIONS

Requirement	Typical Violations
1. Write the original equation.	They write the equation with their value already substituted or a later equation.
2. Substitute your answer in the next line and write a question mark over the equals sign.	No question mark, not substituting for every x, substituting rounded decimal answer for a fractional answer.
3. Simplify each side of the equation, separately in vertically aligned steps, and keeping the question mark over the equals sign until both sides are identical.	Drop question mark early, do not simplify the sides separately, by for example adding a number to both sides. Not vertically aligned steps.

Remedial algebra students violate these steps even with the requirements printed on the tests! Many students from high school have never shown an answer to an algebra problem works. Why is the ability to show your answer works so important?

How can students ever be really sure of an answer? Mathematics is the most epistemologically sound subject. Students should relish the experience and certitude of knowing they are right!

The whole goal of education is to make students independent of teachers, and answers in the back of the book. After all, teachers make mistakes and there are misprints in books, even 10th editions! If a student can show clearly that his answer works in the original equation then he needs no teacher or answer book. This skill comes in very handy on a test!

Showing your answer works is more important than algebraic manipulation. It shows a student the big picture because she makes a complete round trip from unknown x to solving and back to the original equation and her answer working. Showing the answer works makes the student a master of the whole algebraic process!

11. Graphing Problems: Zero, Negatives, Fractions

In elementary algebra, I try to spread the good word of negative numbers and fractions—the students are resistant. I met this resistance in a class devoted to graphing. Graph paper was given and eight function to graph. I walked around helping students individually. Before drawing graphs, they had to make tables of x and y values.. I checked their tables and graphs. Most of the functions were linear, but one ringer thrown in was $y = x^2$. The students were by now used to making tables of x and y values. I helped by emphatically telling them to include negative as well as positive values of x in their table for this function.

So this is why I felt like a missionary among the natives: 1) They would not put negative numbers in their tables. 2) When they did, they squared incorrectly—e.g.

$$x = -2, y = -4 .$$

The lack of negative numbers or the incorrect squaring of them caused the students to draw a line descending from the first into the third quadrant—even after they were warned not to expect a straight line. I also have to be a strong proselytizer of the number zero. Students think somehow that zero is not really a number in good standing. Of course, its use is a great time saver in graphing. When they miss the origin they tend to get shape of this parabola wrong—they miss the curving nature of $y = x^2$. Missing the origin causes another problem. The students who do get the points correct in the second quadrant try to connect those points directly with points in the first quadrant, skipping the part of the curve through the origin.

My next bit of resistance was to fractions. The students were asked to graph $y = \dfrac{1}{x}$. They were told to make an extensive table of x and y values and to be sure to include negative and fractional values of x. As the students worked I gave further hints: Do not expect a straight line. Make x take on many values. This graph is tricky—it is discontinuous, with two parts. Can you guess what these elementary algebra students did?

1. Would not put fractions in for x—same reaction as to negative numbers. Some lamely pointed to the y values that were fractions as satisfying my instructions. They were phobic about writing one over a fraction—possibly for the reason that they were anticipating their difficulty in simplifying if they did.

2. Did not put in nearly enough numbers—had to be forced to put in negative fractions and good-sized integers like $x = \pm 8$.

They were consistent about zero, they ignored it again. Is this a fortunate omission or would it have helped them see how different the graph is?

I conclude from these two graphing experiences that many students in high school need much more experience working with fractions, negative numbers and zero. This lack severely handicaps them in college.

There is one more story from this class. I had noticed weeks before that one male student held his pencil in an extremely awkward way. The pencil was encircled by all four fingers. His hand formed a perfect fist with the end of the pencil emerging below the baby finger. On commenting on his grip he became defensive, "This is the way I write. I write fine." I dropped it quickly—One would expect that an adult of 20 years has faced these comments before and is not going to change easily. Still it was excruciatingly painful to watch him write or draw this way.

But then helping this student graph, an amazing thing happened. I pointed with my finger to one of his plotted points and said it was in the wrong position. He grabbed the pencil in a beautiful position for writing but only to point to the position in question. It was really amazing—his long fingers were elegantly holding the pencil correctly. But to plot the correct position of the point, to actually make a mark, he reverted to the closed fist. It was sad. How can we correct the writing grip of adults?

12. Getting Students to the y=mx+b Form

Elementary algebra students have many difficulties with graphing problems such as:

1) Find the slope intercept form of the line with slope -2 and through the point $(4, -6)$.

2) Find the slope intercept form of the line through the points $(1, -2)$ and $(4,8)$.

3) Find the slope intercept form of the line perpendicular to the line $3x - 4y = 12$ and through the point $(1,2)$.

These problems come after the student has graphed many linear equations and has been introduced to the slope formula:

$$m = \frac{y_2 - y_1}{x_2 - x_1}$$

I often refer to the slope intercept form as the, "y= mx +b" form. The students have been shown how to interpret specific examples of this form. They are shown how to graph a line quickly using the slope and y--intercept.

I have the students do problems like one through three in class while I help individually. Before starting, I assign the studying of solutions in the text to problems of these types. So I almost feel that the students are ready for an in—class work session on these problems, but not quite. I emphasize three things to the students:

a) We are doing the exact reverse of going from equations to graphs. I draw two big circles on the same level. The word "Equations" is put in the left circle and "Straight Lines" in the right circle. I draw an arrow from the left circle to the right and say that we did this already, saying, "To go from equations to graphs you made a table of numbers, plot the points and draw the line." I draw a line from the right circle to the left and say," Today we are doing the exact opposite, going from lines to equations. This is a little harder."

b) If the slope is given, start by writing y = mx +b. I prefer using this form, rather than the point slope form, $y - y_1 = m(x - x_1)$. Next, substitute for m in the next line. If the slope is not given, find it using the formula for slope. Proceed as before. Finally, substitute a known point for x and y and solve for b. Don't forget to write the equation. Teachers with experience know this has to be emphasized.

c) WRITE THE EQUATION with the specific m and b. This really has to be emphasized. I tell them that the answer will have y's and x's. I even give examples of answers:

$$y = 3x + 1$$
$$y = \frac{3}{7}x + 22$$

Admittedly, using the point slope form would avoid the tendency of students to forget writing the equation. I prefer less calculation by starting with y = mx+b. The student needs to realize that the whole point of the exercise is finding the y = mx + b form and should not need a heuristic device to insure he writes the answer.

Here are some problems that the teacher tends to face despite our best efforts to instill the background, knowledge/skill and the pre-work session admonitions above.

1) The students do not solve for b correctly. The difficulties here are various. Not doing the same thing to both sides of the equation. There are problems with fractions and problems with negative signs.

2) NOT writing the equation! Despite strong urging that they must write the equation and even examples of answers, some will stop with the calculation of b. Many have to be shown individually that nowhere on their paper have they written the equation. Often students point to the generic y = mx + b, claiming that they have written it!

3) Problems of type three above cause lots of difficulties. Many books could improve their exercises by the following:

i) Forcing the student to draw the given line, plot the point and roughly sketch the line they are to find.

ii) Not giving points that are so close to the given line-- students think the point given is on the given line.

iii) Many texts have large sets of exercises with common directions above them. Here is an example from a popular text:

For problems 40 through 52 find the slope intercept form of the line with the property given.

40. Through the point (-2,3) and perpendicular to the line 3x-y =7

41. Parallel to the line -2x+4y =8 and through the point (-1,2)

I believe that separating part of the statement of the problem from the other part alienates the geometry and meaning of the problem from the mind of the student. Many students never draw any graphs for these problems—it is possible to get the answer without them. The student should be forced to draw the given line and to plot the given point.

I believe that more variety of problems would be beneficial. Why not find the equation of a line through three points? How about given two parallel lines find two perpendicular lines to them through different points?—Make the student draw the square.

In the next chapter of many Elementary Algebra texts we see systems of equations. This provides the opportunity for a great problem:

Your off road jeep just broke down in the desert at the point $(-2, -4)$. There is a road with equation $3x + 2y = 10$. The units are in miles.

a) Find the equation of the shortest walking route to the road from your location.

b) Find where your walking route will intersect the road.

c) Find the distance you will have to walk exactly and to the nearest tenth of a mile.

d) Find the distance you would have to walk to the road if you went: i) straight north, ii) straight east.

In this problem, the student sees many ideas unified. There is a great benefit in struggling with many ideas in one practical example. In 30 years I have never seen any problem like it in any text from Elementary through College Algebra.

13. The Chicken Way to Do Inequalities and Going the Wrong Way in Equations

The common mistake in solving inequalities is failing to change the sign of the inequality when you multiply or divide by a negative number. I show my students the following example. It shows a solution to an inequality by avoiding the whole issue. I call it the chicken way to solve an inequality.

Solve: $-3x + 1 > 5$

SOLUTION:	WHAT WAS DONE & WHY:
$-3x + 1 > 5$	Original Problem
$-3x > 4$	Minus 1 from both sides
$0 < 3x + 4$	Added 3x to both sides to avoid dealing with a negative 3!
$-4 > 3x$	Minus 4 from both sides
$-4/3 > x$	Divided by 3

The answer is correct. I am not advocating the chicken way. I do it in class (and the normal way) to show an important point: Mathematics is consistent—You get the same answer either way. The following is in the same vein.

GOING THE WRONG WAY IN EQUATIONS

I like to show students that one can do the wrong operation to both sides of an equation and still eventually get the right answer:

Solve: $4x-6 = 9$

My Solution (To illustrate a point):

$4x-6-6 =9-6$ (Subtracting 6 instead of adding!)

$4x-12 = 3$ (Whoops, guess I'll just add 12 now)

$4x-12+12= 3+12$

$4x= 15$

$x = 15/4$

Of course 15/4 is correct, since each equation is equivalent. I like to tell students, "You are not in a straitjacket. If you are careful you'll get the right answer, which is the goal". My goal is to loosen students up. "Relax, there is more than one way to solve an equation (or an inequality). If you are careful you'll get the right answer your way."

14. Mankind's Failure to Understand the Exponential Function

Physics professor Albert Bartlett of the University of Colorado has given a public lecture on population growth and resource depletion for over 40 years on roughly 2000 occasions. He starts every lecture with this sentence. "The greatest shortcoming of the human race is our inability to understand the exponential function." This lecture has been given in 48 states, many foreign countries and videotaped a 1000 times!

I took physics from this charismatic teacher in 1969, the same year he started this crusade, but our paths diverged. I was completely unaware of his obsession until recently. As a mathematics teacher, since 1984, I have independently and in different ways realized that my old professor is correct about mankind and more particularly about math students. Here are two examples showing how students think about exponentials.

Example 1: The very first thing I do with logarithms is to write the particular logarithm:

$$\log_{10}100$$

I tell my students the following two meanings for $\log_{10}100$ and make them memorize exactly:

1. What power of 10 gives 100?"

Or

2. "What do I raise 10 to, to get 100?"

Continuing, I say, "By now you may know the answer to this logarithm which is_____"

It always happens that at least one person says, "Ten." From this I realize that they are confusing multiplying with exponentiation. They are confusing 10 times 10 with 10^{10}.

This is quite striking— I make sure to say the two sentences slowly and pause before soliciting their answer. The result is always the same—several people say ten. This is symptomatic of people's failure to understand the exponential function. Now, for my second example.

A well-known riddle goes like this. In a jar there is one jelly bean. Every minute the number of jelly beans doubles. After 30 minutes the jar is full. When was the jar half full? I am careful to admonish those who have heard this riddle to keep mum. Only one or two people in College Algebra classes figure this one out right away. I can understand this but what is quite striking is how stubbornly people will stick to the answer 15 minutes or will not see that 29 minutes is the answer, even after the reasoning is given.

Here is one way to show that the exponential function is different. I setup a race between functions as follows:

RACE BETWEEN FUNCTIONS				
x	5x	x^2	x^5	2^x
4	20	16	1024	16
10	50	100	100,000	1024
20	100	400	3.2 million	1.05 million
30	150	900	24.3 million	1.07 billion
40	200	1600	102 million	1.1 trillion

I fill in the table by rows and talk like I'm narrating a race. At x equal to 4, the power function, x^5 has a big lead with x^2 and 2^x tied for second place. At the second row, draw attention to x^5 , still strongly in first place at an impressive 100,000, but the exponential function 2^x has claimed second place and left the quadratic x^2 far behind. In row three or x equal to 20, I say, " x^5 is still leading, but 2^x is in the same ballpark, and both are way ahead of the others." By x equal to 40 the exponential 2^x is more than 10,000 times as much as x^5 and will keep increasing its ratio to x^5 as x increases.

Of course, graphs of $y = 2^x$ compared to $y = x^5$ also show how different these types of functions are.

I close with one of professor's Bartlett's favorite stories. Suppose there were some bacteria in a Petri dish. A biologist has provided plenty of nutrients for them, and they happily double in numbers every hour. After 12 hours (twelve generations for these little fellows), the bacteria are scattered widely over the dish, but it is only one-eighth full. At this point, one of the more intelligent bacteria convenes an, "All Dish Population Convention." He rises to an imposing height for a bacterium and states in an authoritative voice:

"My fellow germs, as you know, we have been doubling every hour, and our beautiful home is now one-eighth full. At this rate, our dish will be completely full in only three more hours. That is only three generations from now! We must take steps or mass overcrowding and starvation will surely result.

This mathematically intelligent plea falls on deaf ears. The audience of bacteria say, "Look, we have been doubling for twelve generations and that is a very long time. Our dish is pretty empty. There are still 7/8 ths of the dish completely unoccupied. Why should we worry?"

Substituting people for bacteria, earth for Petri dish, and 23 years for one hour should be enough to scare anyone. It is mankind's failure that it does not.

15. Logarithms and Ratio Growth

The following property of logarithms is well known, but neglected in teaching. Consider a quantity that is growing by constant ratio jumps. The logarithms of the quantities will differ by constant increments. This is true no matter what number you start with and no matter constant ratio you pick.

I like to illustrate this dramatically with a classroom exercise. Ask the class to pick any positive integer. Suppose they pick 7. Ask them for another positive integer to multiply it by. Suppose they pick 4. Keep multiplying and start a table on the board as follows.

Initial x, &Keep Multiplying by Four	Log(x)	Difference from Last
7		
28		
112		
448		
1792		

Have the students find all the logs for the second column, rounded to the nearest tenth. Then fill in the last column, showing the differences in the previous logarithm. Look for patterns. The final table is:

x	Log(x)	Difference from Last
7	.8	--------
28	1.4	.6
112	2.0	.6
448	2.7	.7
1792	3.3	.6

We see that the differences in the logs are always either .6 (or .7 due to rounding). Note that the starting integer and the multiplier are both chosen by the audience, yet this pattern always happens—pretty striking. Emphasize also that the differences in x are not staying the same.

So why do logs turn ratio growth into difference growth? The reason is simple. Should we explain why in class or make the students derive this property? Here is an argument.

Let: x = starting integer, and m = multiplier or the constant ratio

The entries are: $x, mx, m^2x, m^3x,...$

But: log (mx) = log (m) + log(x),

The first difference of the logs is then log(m). Ask the students how to extend the argument to log(m^2 x) versus log(m x).

In our case, log(m) = log(4) $\cong$ .602

Be sure to summarize in words for the students: "Constant ratio growth is turned into constant difference growth by logarithms."

Now, for some physiology. Over a thousand years ago, Arabs classified the stars into brightness magnitudes.

These magnitudes, such as -6, -5, -4 ... are still used, but now we have electronic instruments which measure the photon energy reaching us from stars. These instruments show that each step in magnitude classification, which remember came from human eyes 1000 years ago, differs from a lower step by the constant ratio of about 2.5 times as much in energy. Thus, human eyes act like logarithms: Their eyes converted constant ratios of light energies to constant differences on a scale we call brightness magnitude. There are many other examples of this human sensing property.

My favorite example of this sensing property is in the design of the piano. In the middle is a white key, called, appropriately, Middle C. Striking this key sets a wire vibrating at about 250 hertz. The next higher C note is 12 keys to the right (one octave higher). This High C key sets a wire vibrating at about 500 hertz. What about the next higher C, another 12 keys to the right? Ask the students! It vibrates at about 1000 hertz. Why not 750? It would not sound right. If you want the 'difference' in pitch between C and High C to sound the same as the difference between High C and the next higher C you have to make the ratio of the frequencies two.

I joke that we should have a Martin (my name) scale of pitch defined as follows.

M = log (f/250), where f is the frequency

So on this scale, soon to be adopted worldwide, frequency ratios would be turned into constant differences.

There are many other examples of log scales, having nothing to do with physiology, such as pH in chemistry or the Richter scale. But physiology examples are fun.

One more! Experiments with people have shown that they can almost always tell that 11 pounds is heavier than 10 pounds. But they frequently confuse the heavier of 50 and 51 pounds. Ask your students what is going on here. How heavy should the heavier weight be so that it is distinguishable from the 50 pound weight?--Assume weight sensing is determined by ratios. The sensing of light, sound frequency and many other things depend on ratios, not differences.

16. Teaching Variation

Our math texts emphasize taking variation statements and translating them into equations. For example:

"The surface area of a sphere varies directly as the square of the radius. "The translation to mathematics is, of course:

$$A = kr^2$$

Translating the other way, from a formula to an English statement is hardly ever mentioned. It is just as important. Here is an example.

"Make a statement of variation for the following which deals with the power output P of an electric windmill driven by a wind of speed s". Suppose the specific formula is:

$$P = 5s^3$$

The statement of variation (or translation to English) is: "The power of a windmill varies directly as the cube of the wind speed."

Students frequently say the constant 5, which is not desired. We will see that putting constants in the translation to English can obscure relationships.

Do students realize the impact of the cube power? Ask them what happens to the power if the wind speed doubles. Students have a lot of trouble with this.

To answer this last question, I emphasize that forming ratios is an important skill—it is especially useful in variation. Let the subscripts H and L stand for a higher and lower wind speed respectively. I write the following on the board.

Since the wind speed doubles, we have:

$$\frac{P_H}{P_L} = \frac{5(2s)^3}{5s^3} = \frac{8s^3}{s^3} = 8$$

We see that the power is eight times as great. Many students do not have the initiative, good notation, and algebraic skills to do this. We see that constants, such as 5 in this windmill relationship, are an impediment to clear thinking when ratios are considered.

Here is another important and rarely covered aspect of variation. Can the student take a relation based on the radius of circle:

$$A = \pi r^2$$

and get a formula relating area and the diameter? Of course they are expected to use $d = 2r$. There are very few problems in texts requiring the manipulation of two or more formulas at once. Physics students are always doing it!

I close with three favorite variation problems which show the power of this subject.

1) The load that a beam can support varies directly with the width and the square of the depth. A roof beam is a 2 by 4 and supports a weight with the longer dimension vertical. How many times greater or less weight can be supported if the beam is rotated 90 degrees so that the longer dimension is horizontal?

2) A scientist wants to study the change in the length of a piece of steel due to temperature increase. She conducts experiments which are summarized in the table:

Original Length	Change in Temp.	Change in Length
10 meters	50 degrees	.2 millimeters
10 meters	100 degrees	.4 millimeters
20 meters	100 degrees	.8 millimeters

a) The first two lines of data show that a doubling of the temperature change caused a doubling of the change in length. This is the hallmark of what simple type of variation? Formulate a variation statement relating change in length with temperature change but ignoring the original length. Write its mathematical equivalent.

b)   Now compare just the second line and the third line. We see a doubling in the change in length for the same temperature change! Thus there is a second factor causing the length to change. Formulate in words a more complete statement of variation, relating the change in length to both the original length and the temperature change. Write its mathematical equivalent.

Answer: "The change in length varies jointly as the temperature change and the original length. "Mathematically:

:

$$\Delta L = k(\Delta T)L$$

My last example has three parts.

3) Johannes Kepler stated in 1619 that, "The square of the time for any planet to orbit the sun varies directly as the cube of the mean distance from the sun."

a) The earth takes one year to orbit and its mean distance from the sun is one Astronomical Unit (AU). Translate into mathematics and show that k equals one. This is Kepler's law for the orbital period. Pluto was discovered in 1930, when a very faint object was detected moving on a series of photographic plates. Is the image faint because it was very far away, very small, or what combination of these factors? From the movement on the pictures they determined that it would take 248 years to orbit the sun. Now you can find the mean distance from Pluto to the sun using Kepler's law.

b) A new asteroid is discovered with a perihelion of .5 AU and aphelion of 1.3 AU. Find the time for the asteroid to orbit the sun. Note: For all objects orbiting the sun, the perihelion equals $(a - c)$, where a is the mean distance from the sun, used in Kepler's law. The c is the distance to the focus from the center of the ellipse. The aphelion equals $(a + c)$. Thus, solving a simple system of equations gives both a and c, the ellipse parameters.

c) Find the equation of the elliptical orbit of the asteroid using:

$$\frac{x^2}{a^2} + \frac{y^2}{b^2} = 1$$

Where $b = \sqrt{a^2 - c^2}$

Isn't it amazing that from just two numbers, the aphelion and perihelion, we can find the equation of a planet's orbit and the time it takes to orbit?

17. Artificial Motion

Every algebra book has the iconic motion problems we see over and over. "A car is going east at 60 mph and another west at 50 mph. When will they be 300 miles apart?" Or: "A boat is on a river. If the current is ... ." All of these problems have motion at constant speed—there is never, and I mean never, acceleration. But no motion exists on this earth without acceleration! Every runner, car train, and plane, starts and stops! By college, shouldn't there be an attempt to match reality? I argue in this section that acceleration problems should be in every math class from elementary algebra to college algebra and definitely before calculus.

Some would say, "Acceleration is physics-- we have enough to do in mathematics."

This attitude does four things:

1) It cuts off 70% of the population (The approximate percent who never take physics or calculus) from the real world and gives them a false sense that they are educated. Everyone sees and experiences acceleration every day!

2) It eliminates a beautiful mathematical subject. Suppose an object is moving with constant acceleration. A graph of velocity on the vertical versus time on the horizontal axis is a line with constant slope. The area under this line is the distance covered. This is math as much as it is physics. The calculus people are not 'protected' from this reality, but are expected to understand it with no background in acceleration. Algebra students should be working with acceleration, as no calculus is needed to calculate the area (distance) under a straight line. It's either a triangle or a trapezoid.

Working with a graph like this, is one of the best

contexts in which to show two quantities related linearly—

movement is a very natural context. This context is more

natural than cost functions or the hundred other applications

of linear functions in texts.

3) It ignores an application of mathematics to the

physical world which led to an explosive development of

mathematics in modern times. Galileo used a graph to figure

out the distance covered by an accelerating object. How far

does an object go starting at rest and accelerating at a

constant rate for a fixed time? His genius was to consider

another object moving at a constant speed equal to half the

final velocity of the accelerating one. Galileo argued that the

distance covered by both objects are the same if they travel

for the same time. He convinced himself (and Western

Europe!) of this equivalence by geometry. He actually put a

graph of velocity versus time for both objects (400 years ago!)

in his book, "Discourses and Mathematical Demonstrations

Relating to Two New Sciences." This book was banned by the Roman Inquisition. Its ideas seemed to be banned in all math textbooks below calculus. Also note that Galileo used the phrase, 'Mathematical Demonstrations' in the title. It's mathematics, not just a physics subject! The strong impetus of acceleration to mathematical discovery continues with the revered Newton. He was motivated by the acceleration of falling bodies to develop calculus.

4) No acceleration means you can't talk intelligently about force (Force = Mass × Acceleration). Again, this cuts people off from the physical world.

Calculus students need to understand acceleration, because it provides the best example of the need for derivatives--finding instantaneous speed. For a calculus student with no physics, acceleration feels like it is suddenly sprung on her. She is absorbed by many things: limits, algebraic techniques, and derivative rules. Anti-derivatives are coming soon in the course. Acceleration has to be understood quickly and in passing. By calculus, mathematics teachers have to confess, "Whoops, acceleration is very important. When an object's acceleration is not constant, you need to do an integral to find the velocity function. Oh, and then another integral of the velocity function to find the distance covered." How understandable is that if you've never had problems with constant acceleration?

A student successfully completing the three semester sequence of Elementary, Intermediate, and College Algebra is Aristotelian in their ignorance of motion and what causes it.

18. Euler's Totient Function

The function notation, f(x), is rightly emphasized in algebra books but the advantages of its use are not seen in typical examples such as graphing f(x) = 2x-1.

Why is this any better than y = 2x-1?

I suggest Euler's Totient (From the Latin, meaning "so many") function. It is usually written as $\phi(n)$ and equals the number of integers less than n that are relatively prime to n. The number one is always counted as relatively prime to n, by convention.

I give my students the table:

n	Numbers Below n and Relatively Prime to n	$\phi(n)$ (=Count of Numbers in 2nd Column)
1	{1}-Note: One is always counted	1
2	{1}	1
3	{1,2}	2
4	1,3}	2
5	{1,2,3,4}	4
6	{1,5}	2
7	{1,2,3,4,5,6}	6
8	{1,3,5,7}	4

Then give the following exercises:

1) Extend the table to n =20.

2) Fill in the blanks: If p is a prime then complete the following:

$$\phi(p) = \underline{\hspace{3cm}}$$

3) If n is a power of a prime, say p^k then

$$\phi(p^k) = p^k - p^{k-1}$$

Give two examples from your table.

4) If a and b are relatively prime, then

$$\phi(ab) = \phi(a)\phi(b)$$

Using this rule find $\phi(45)$ *and* $\phi(95)$. In both cases, list all the numbers that are relatively prime as a check.

5) Find $\phi(1000)$ *and* $\phi(53000)$ using combinations of the rules above.

I have found that many students use the function notation incorrectly in these exercises. They write:

$\phi(45) = 5 \times 9 = 4 \times 6 = 24$, when the second step should be

$\phi(5) * \phi(9)$. The multiplicative nature of this rule is succinctly

expressed with the function notation, and this is a context of

the use of function notation that students should see.

Playing with Euler's Totient Function

There are many other things to do with Euler's

function. Number theory is so neglected in lower level

mathematics but I try to make up for that here and later in

this book.

We can look at the ratio $\phi(n)$ / n. This ratio is a

measure of the 'primeness' of n, but scaled to the size of n.

Have the students search as follows:

For numbers less than or equal to 100:

a) What types of numbers n, have a high value of $\phi(n)$ /n?

Which has the highest?

b) What numbers have the lowest ratio $\phi(n)$ /n? Hint: There is a three way tie! What prime factorizations do the three-way tie numbers have? Find a number over 100 that has a lower ratio.

c) What numbers have a ratio $\phi(n)$ /n of $1/2$?, of $1/3$? What prime factorizations do they have?

Part II: College Algebra, Trigonometry, and Calculus

1. A Rational Function To Open Their Eyes

I suggest this graphing problem with its unusual interaction with an asymptote:

"Find the asymptotes and graph the function

$$y = \frac{x^3 - x^2 - 6x}{x^2 - 3x + 2}$$ "

It factors as

$$y = \frac{x(x-3)(x+2)}{(x-2)(x-1)}$$

We see that x = 1 and x = 2 are vertical asymptotes. There is an oblique (or slant) asymptote because the degree of the numerator is one more than the denominator. It is found by polynomial division and is y = x+2. Notice that (x+2) is also a factor in the numerator. This will have an interesting effect.

A student, using the standard graphing window of (x and y ranging from -10 to 10), might see this:

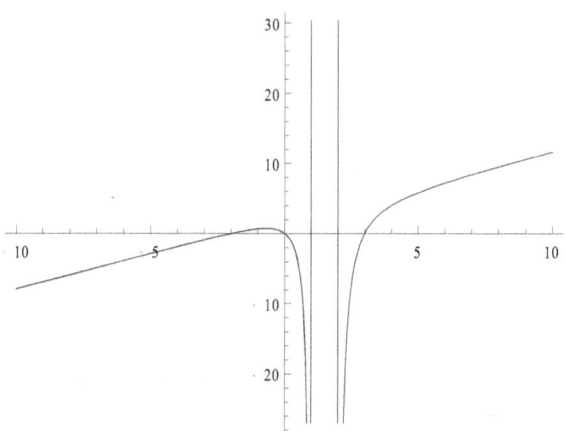

There are two errors here. First, the vertical asymptotes have been plotted. Many calculators, such as the Texas Instruments family, have corrected this artifact. Another, more serious error is that the portion between x = 1 and x=2 is missing. Increasing the y range gives that part: in the graph below.

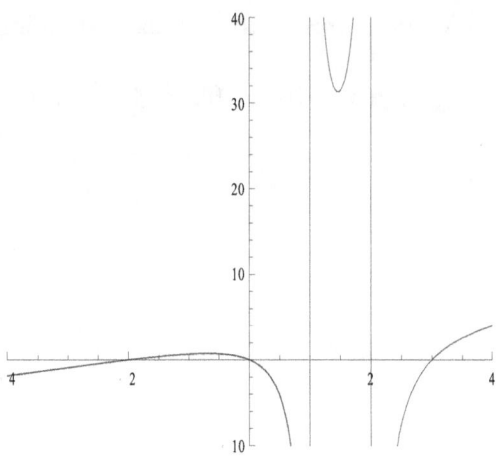

These graphs were done with Mathematica. It insists on graphing the vertical asymptotes. Finally, let's concentrate on the region around $x = -2$ and plot the slant asymptote, $y = x+2$. See next graph. As x decreases from zero to -2, the function is below $y = x+2$ and approaching it from below. At $x = -2$, they intersect. The function and its asymptote have the common point, $(-2,0)$. For x decreasing below -2 the function is now above the asymptote, but asymptotically approaches it from above forever as x goes to negative infinity.

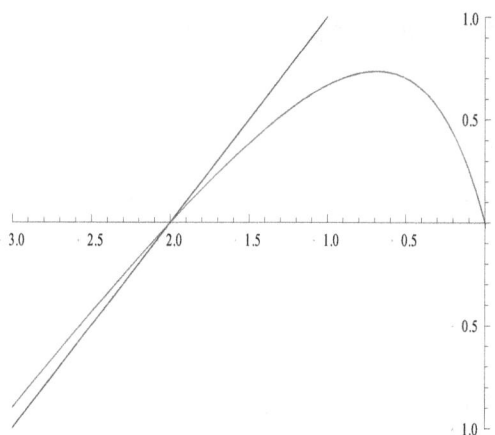

The coincidence that (x+2) is a factor in the numerator and also the slant asymptote ensures the intersection of the two graphs.

2. Three Models of Growth

College Algebra students see three models of growth:

Math Model—Name	Usual Context, Application
(1) $A = p(1+r)^t$—Discrete Compounding	Money in a Bank
(2) $A = p\,2^{(t/D)}$—Doubling Model	Population Growth
(3) $A = pe^{kt}$—Continuous Growth	Population Growth or Isotope Decay

It is unfortunate that the contexts for the models are rarely varied. Model (1) is rarely used in the context of isotope decay (r negative). Model (2) is usually not used in compounding interest.

But all three models can be used in each of the three contexts. I like to show students this and to numerically compare the models. In the example below, I will assume that model (2) is exact. Then I will find approximate parameters for the other two models, and finally compare the predictions over a long time period.

Suppose the application is a population that has an exact doubling time of 25 years. Then three models giving close (or exact in the (2) case) values for this situation are:

$$(1') \ A = p \ (1+.0281)^t,$$

$$(2') \ A = p2^{(t/25)}$$

$$(3') \ A = pe^{.0278t}$$

Formulas $(1')$ and $(3')$ use rates rounded to the nearest thousandth. In all cases, t is the time in years. In the next table I compare the models over the time range 25 years to 800 years.

Comparison of 3 Math Models—For a

Population With an Exact Doubling Time of 25 Years

t= # of Years	$A=p2^{(t/25)}$ (Exact)	$A=p(1+.0281)$, with Rel. Error	$A=pe^{.0278t}$, with Rel. Error
25	2p	1.99933p,.03%	2.0037p,.19%
50	4p	3.997p,. 07%	4.01p, .37%
100	16p	15.98p, .1%	16.119p,.74%
200	256p	255p, .2%	260p, 1.5%
400	65536p	65184p, .5%	67508p,3.0%
800	4294967296p	$4.249*10^9$p, 1.1%	$4.56*10^9$p,6%

The relative errors are based on the doubling time of exactly 25 years model. It is interesting that for all time values and for the same rounding to 3 decimal places the continuous model gives a relative error that is 5 to 7 times the relative error of model one, the annual compounding model.

This can be understood by comparing the relative errors for one year in both models. For the same decimal place rounding the relative error in the annual compounding was only .05% but for the continuous case the relative error was .27% . The ratio of these rounding errors is just over 5, which is within the 5 to 7 times ratio stated.

3. Population Mathematics at Different Levels

The population of the USA and how it has changed from earliest times should be of interest to students. This section has three examples of mathematics applied to populations at different levels of maturity.

I .Mathematical Level: Arithmetic and Elementary Algebra

The United States was the first country in modern times to require a regular census. I present the 10 year numbers going way back to 1790 and ask them (and you) to fill in the last column:

Year	Pop. in Millions	Rel. Change vs. Previous Census
1790	3.9	(Not Applicable)
1800	5.3	
1810	7.2	
1820	9.6	
1830	12.9	
1840	17.1	
1850	23.2	
1860	31.4	
1870	39.8	
1880	50.2	
1890	62.9	
1900	76.2	
1910	92.2	
1920	106.0	
1930	122.8	
1940	132.2	
1950	150.7	
1960	179.3	

1970	203.3	
1980	226.5	
1990	248.7	
2000	281.4	
2010	308.7	

What 10 year period was the first to have less than 30% growth? What are some possible reasons? (Make them give historical reasons) What 10 year period had the lowest 10 year rate? Why? Many people guess that the 1940's containing WWII had the lowest relative change—It was not. The lowest decade of relative change was the 1930's. Ask them why. There is a lot of history in these numbers and social changes! How many children did the average woman have in 1790?, 1890? , or 1990? When did birth control come in and how did it affect the population growth? Most almanacs have the area of the USA every census year and the population per square mile. It is fun see how these have changed. Also the center of population has moved. Ask how they are all calculated.

II. Mathematical Level: Intermediate Algebra and College Algebra

The populations in millions of India I and China C can be modeled respectively as

$$I = 800 \ e^{.015t} \quad \text{and} \quad C = 1000 \ e^{.0071t}$$

respectively, where t is the number of years after the year 2000. With India's 1.5% growth rate versus China's .7%, India is sure to eventually equal China's population, but when? Students are eager to put both functions on graphic calculators and approximate the intersection. They have trouble finding the solution analytically:

$$I = C$$

$$800 \ e^{.015t} = 1000 \ e^{.0071t}$$

$$e^{.015t} / e^{.0071t} = 1000/800$$

$$e^{.008t} = 1.25$$

$$t = \frac{ln\ 1.25}{.008} \cong 28 \text{ years}$$

III. Mathematical Level: College Algebra and Precalculus

We can use a stochastic matrix to model the fractions of a populations suffering from a disease over time. Suppose the 4 fractions of the population are: Uninfected = U, Infected= I, Cured and immune= C, and Dead= D.

Suppose the transition probabilities are :

	U	I	C	D
U	.7	0	0	0
I	.3	.25	0	0
C	0	.6	1	0
D	0	.15	0	1

The entries under the column headings give the probabilities of transitioning from that category to the row category in the next time period. For example looking at the column under the letter U, the transition matrix is understood as follows. If one is uninfected (U), the probability of staying uninfected (The U row) is .7 through the next time period.

Similarly the probability of an uninfected person transitioning to infected is .3. The U column also implies that an uninfected person cannot become cured or dead in the next time period, denoted with the two zeros at the end of the U column. The cured column, C indicates that all cured people stay cured and similarly for the dead column. (Implied by the sole entry of one in those columns). The transition time period between states could be one month. From these 16 numbers we get the 4 by 4 transition matrix T:

$$
\begin{bmatrix}
.7 & 0 & 0 & 0 \\
.3 & .25 & 0 & 0 \\
0 & .6 & 1 & 0 \\
0 & .15 & 0 & 1
\end{bmatrix}
$$

Let the column matrix S, (for state) be :

$$
\begin{bmatrix}
u \\
i \\
c \\
d
\end{bmatrix}
$$

where the lower case letters represent the initial fractions of uninfected, infected, cured and dead respectively from top to bottom.

We must have:

$$u + i + c + d = 1$$

The product $T\,S$ gives the four fractions of the population after one month. T^2S are the fractions after two months etc. One way to find the final stable state of the population is to take the matrix T to a high power. For example, T^{40} (40 months) on a graphing calculator gives:

$$\begin{bmatrix} .0000006 & 0 & 0 & 0 \\ .0000004 & 8*10^{-25} & 0 & 0 \\ .8 & .8 & 1 & 0 \\ .2 & .2 & 0 & 1 \end{bmatrix}$$

We may assume the values very close to zero will go to zero in the limit of infinitely many months. Put zero in for those three entries. Then multiply that matrix with an initial population distribution with the fractions u, i, c, and d

corresponding to the initial fractions that are: uninfected,

infected, cured and dead respectively, i.e. with the column

matrix:

$$\begin{bmatrix} u \\ i \\ c \\ d \end{bmatrix}$$

The result is the 4 by 1 column matrix:

$$\begin{bmatrix} 0 \\ 0 \\ .8\,u \,+\, .8i \,+\, c \\ .2u \,+\, .2i \,+\, d \end{bmatrix}$$

This general result, for any initial fraction distribution

of people, gives the final distributions in the states U, I, C and

D from top to bottom respectively. The first two zero entries

imply that no one is left in the uninfected or infected states.

The third row is the cured category of people. Its entry means

that 80 % of the original uninfected and 80 % of the original infected will eventually become cured. One hundred percent of the original cured will stay cured by assumption and indicated by the c in the third row. The fourth row (dead) entries imply, similarly, that 20% of the uninfected and 20% of the infected will transition to dead. Of course 100% of the original dead will stay dead. It is very interesting how this 80%—20% split is inherent in the original transition matrix— we don't see .8 or .2 anywhere there! I close with following story from a differential equations class, studying applications to populations.

We were studying the equation:

$$\frac{dP}{dt} = bP - dP$$

P is the population as a function of time t. The variables b and d are the birth and death rates respectively.

For data we looked at the relative change in different 10 year periods of the US census, given above. The relative increase calculated from 1980 to 1990 stands out as only 9.8

%. The previous three 10 year changes were 11%,13% and

12%. The change from 1990 to 2000 was back up to 13.1 % I

asked the class, "What happened in the 80's? A student who

hardly ever answered questions responded with no hesitation,

"H M O's!"

4. Highly Divisible Numbers (Part I)

A nice application of the Fundamental Principle of Counting is a formula for the number of divisors that an integer has. The word number will refer to a positive integer. Let a number n have prime factors: $p_1, p_2, ... p_k$. If the prime factorization of n is given as:

$$n = p_1^{e_1} p_2^{e_2} p_3^{e_3} ... p_k^{e_k}$$

Then the number N of divisors n has is given by

$$N = (1 + e_1)(1 + e_2)(1 + e_3)...(1 + e_k)$$

For example $12 = 2^2 * 3^1$

The number of divisors of 12 is $(2+1)(1+1) = 6$.

The six divisors are 1,2,3,4,6 and 12.

Note that our formula always includes one and the number itself. Here are some problems for students.

Problem Set One, Answers Follow

1. What numbers have an odd number of divisors? Give an example.

2.a) What positive integer less than 50 has the most divisors?

2.b) For numbers less than or equal to 100, there is a five way tie for the most number of divisors! Find the five numbers and the number of divisors they have.

Answers

1. Only numbers with prime factorizations in which all exponents are even have an odd number of divisors. An example is $2^4 * 5^2$. There are $(4+1)(2+1)$ equals 15 divisors.

2. a) 48 has the most with 10 divisors.

b) 60,72,84,90 and 96 all have 12 divisors.

A Measure of Divisibility

It's easy to create a number with many divisors. Just use many prime factors with exponents that are large. A more interesting challenge is find a number that has many divisors for its size. For example 60 has 12 divisors. This is at least 4 more divisors than any of the numbers from 49 to 71. Let's try to devise a scale of divisibility that takes into account the size of the number. We could try S(n) defined as follows:

$$S(n) = \frac{\text{number of divisors of n}}{n} = \frac{N}{n}$$

However, this measure of divisibility is not very interesting. It is highest at the two lowest positive integers! :

$$S(1) = 1, \quad S(2) = 1,$$

For example, S(3)= 2/3 and S(4) is .75. S(n) generally decreases as shown in the following plot.

S(n) Versus n for n = 1 Through n = 50

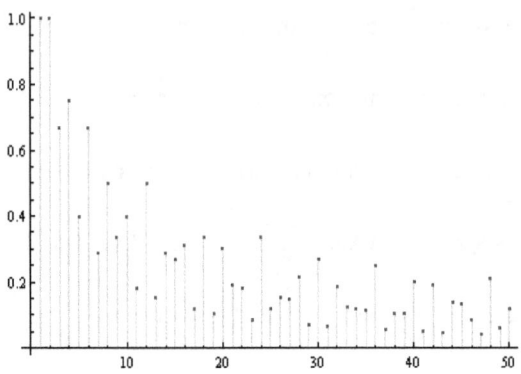

<u>Problem Set Two for Students--Answers Follow</u>

1. Discuss S(p) for p a prime bigger than 2. Find a formula for S(p). As p increases, what happens to S(p)?

2. Let n be power of 2, say $n = 2^e$. Discuss $S(2^e)$ as e increases without limit.

3. Now consider n a power of any prime p. Let $n = p^e$. Show that $S(n = p^e)$ is a maximum at e = 1.

4. Do problems one through three imply that S(n) is a maximum for numbers that are equal to one or two?

Answers

1. $S(p) = \dfrac{2}{p}$. For all primes p. We see that S(p) is less than one for primes three or greater and goes to zero with increasing p.

2. $S(2^e) = \dfrac{e+1}{2^e}$. It is easy to show that as e increases, $S(2^e)$ decreases. That is, $S(2^e)$ is a decreasing function. Graph it or take the derivative. Thus two itself has a greater divisibility than for any higher power of 2.

3. $S(p^e) = \dfrac{e+1}{p^e}$. Again, it is easy to see that this is a decreasing function of e as e increases.

4. No. One has to consider numbers that are products of primes and/or products of primes raised to powers. But it is fairly easy using the arguments above to see that S(n) is a less than one for numbers such as these.

A More Interesting Measure of Divisibility

An interesting measure of divisibility, D(n) is defined as follows.

$$D(n) = \frac{Number\ of\ Divisors}{\sqrt{n}}$$

Here is a graph of D(n) versus n. What do you notice?

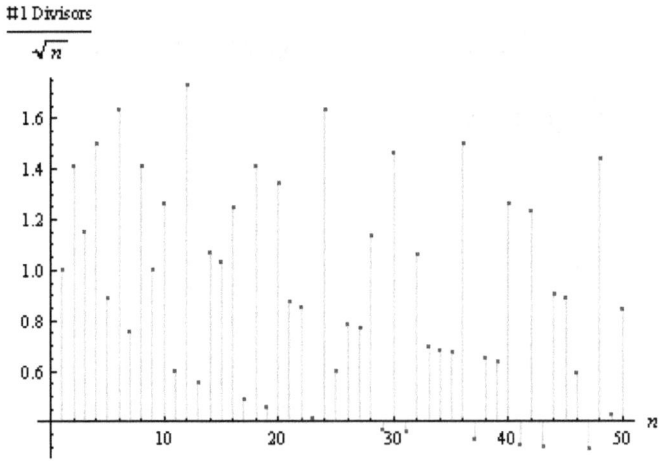

We see, with this arbitrary measure of divisibility,

that 12 has the highest divisibility for numbers in the range 1 to 50. In fact:

$$D(12) = 6 / \sqrt{12} = \sqrt{3}$$

Also 24, 36 and 48 are notable standouts. Does any integer beat 12's divisibility of $\sqrt{3}$? I think the answer is no. The proof seems to take some effort.

5. Highly Divisible Numbers (Part II)

What makes an integer number highly divisible? Six is a perfect number but 12 is the first integer, the sum of whose proper divisors exceeds itself. This, by the way, would make 12 an excellent base for frequently occurring fractions such as $1/2, 1/3, 1/4$ and $1/6$ as they would terminate in a base 12 system.

In this section, I look at the sum of the proper divisors of a number as a measure of divisibility. In the previous section, the measure was based on the number of divisors. So what makes the sum of the proper divisors large? Systemic observation is the first step in science. I make students construct tables like the following.

SUM OF PROPER DIVISORS for NUMBERS TWO through TWENTY		
Number	Sum of Proper Divisors	Prime Factorization
2	1	prime
3	1	prime
4	3	2^2
5	1	prime
6	6	2*3
7	1	prime
8	6	2^3
9	4	3^2
10	8	2*5
11	1	prime
12	16	$2^2 * 3$
13	1	prime
14	10	2*7
15	9	3*5

16	15	2^4
17	1	prime
18	17	$2*3^2$
19	1	prime
20	22	2^2*5

Are there rules we can get from this table? All primes have sum one, pretty trivial! There are trends: Composites with different divisors (e.g. 6, 12 and 18) have high sums. The sum of divisors can be made arbitrarily large—just make a number with a lot of factors. As in the previous chapter, it is more interesting to create an arbitrary scale of divisibility based somehow on the size of the number. Suppose we define d(n) as follows.

$$d(n) = \frac{\text{sum of proper divisors}}{n}$$

We will call d(n) the divisibility (our latest definition of this word!) of the number n. You can check that the number with the highest divisibility in the table is 12, with

d(12) = 16/12 =4/3 . The game is then finding an n for

which d(n) is as large as possible.

The sum of the proper divisors of an integer is again

based on the prime factorization. Suppose the prime

factorization of n is given by:

$$n = p_1^{e_1} p_2^{e_2} p_3^{e_3} ... p_k^{e_k}$$

The sum S of the proper divisors of n is given by

$$S = (1 + p_1 + p_1^2 + ... p_1^{e_1})(1 + p_2 + p_2^2 + .. p_2^{e_2})..(1 + p_k + p_k^2 + ... p_k^{e_k}) - n$$

$$= \frac{1 - p_1^{e_1+1}}{1 - p_1} \frac{1 - p_2^{e_2+1}}{1 - p_2} \frac{1 - p_3^{e_3+1}}{1 - p_3} ... \frac{1 - p_k^{e_k+1}}{1 - p_k} - n$$

So:

$$d(n) = \frac{S}{n} =$$

$$\frac{p_1 - p_1^{-e_1}}{p_1 - 1} \frac{p_2 - p_2^{-e_2}}{p_2 - 1} \frac{p_3 - p_3^{-e_3}}{p_3 - 1} ... \frac{p_k - p_k^{-e_k}}{p_k - 1} - 1$$

1. What is the upper limit of d(n) when n is equal to an

   arbitrarily large power of 2?

2. What is the upper limit of d(n) when n equals products of

   2 and 3, both to arbitrarily large powers?

3. a) Suppose n is a product of the form:

$$n = p_1^x p_2^x p_3^x \cdots p_k^x$$

Discuss whether or not the following limit exists.

$$\lim_{x \to \infty} d(n) = \lim_{x \to \infty} d(p_1^x p_2^x p_3^x \cdots p_k^x)$$

b) Give an example.

Answers

1. One, for example

$$d(2^{30}) = \frac{2 - 2^{-30}}{2-1} - 1 \cong .99999999919$$

2. Two, for example

$$d(2^{20}3^{15}) = \frac{2 - 2^{-20}}{2-1} \frac{3 - 3^{-15}}{3-1} - 1 \cong 1.9999985$$

3. a) $\displaystyle \lim_{x \to \infty} d(n) = \frac{p_1}{p_1 - 1} \frac{p_2}{p_2 - 1} \frac{p_3}{p_3 - 1} \cdots \frac{p_k}{p_k - 1} - 1$

It follows from problem 3 that if allowed to range over all primes, d(n) can be made arbitrarily large. The reason is that all factors of the form:

$$\frac{prime}{prime - 1}$$

are bigger than one and of course there are infinitely many primes. It's amazing how often that fact is crucial.

We see that d(n), our arbitrary measure of divisibility, does not result in a unique integer with greatest divisibility—d(n) can be made arbitrarily large.

One final question. Which n gives a higher d(n):

$$n_1 = 2^{10} * 3^{10} * 5^{10} \quad \text{or} \quad n_2 = 2*3*5*7*11*13 \text{ ?}$$

In general, is d(n) made larger by having fewer primes to high powers or by having more primes?

I leave it to the reader to explore this vaguely worded question.

6. The Number of Prime Factors Dividing and Not Dividing Into all the Numbers Less than a Number

Suppose a positive integer n has k prime factors: $p_1, p_2, p_3, ..., p_k$. Let us form these primes into two groups as follows.

$$p_{d1}, p_{d2}, ...p_{dD} \text{ and } p_{n1}, p_{n2}, ...p_{nN}$$

where D and N are integers with $0 \leq D + N \leq k$ and $D, N \geq 0$

The primes in the first group (with subscript D) will be called the divisible group. The primes in the second group will be called the non-divisible group.

There is a very nice function of n which tells how many numbers not exceeding n satisfy the following two conditions:

1. Are divisible by the primes in the divisible group:

$p_{d1}, p_{d2}, ... p_{dD}$ and may be divisible by other prime divisors

of n, but:

2. Are not divisible by the primes in the non-divisible group:

$p_{n1}, p_{n2}, ... p_{nN}$

The function, denoted $n_{\{p_{d1}, p_{d2}, ... p_{dD}\}, \{p_{n1}, p_{n2}, ... p_{nN}\}}$, which tells

how many numbers satisfying both conditions is given by:

$$n_{\{p_{d1}, p_{d2}, ... p_{dD}\}, \{p_{n1}, p_{n2}, ... p_{nN}\}}$$
$$= \frac{n}{(p_{d1} p_{d2} \bullet \bullet \bullet p_{dD})} (1 - \frac{1}{p_{n1}})(1 - \frac{1}{p_{n2}})...(1 - \frac{1}{p_{nN}})$$

By convention we define: $n_{\phi, \phi} = n$

Here are some examples with listings of the numbers

satisfying the conditions to check.

$12_{\{2\},\{3\}}$ would be the number
of integers not exceeding 12
that are divisible by 2 but not by three.

Thus, by our formula:

$$12_{\{2\},\{3\}} = \frac{12}{2}(1-\frac{1}{3}) = 4.$$

Checking, the numbers not exceeding 12 that are
divisible by two but not by 3 are:
2,4,8,10. There are 4 numbers as predicted.

Let's put 3 in the divisible group and 2 in the
non-divisible. Thus,

$$12_{\{3\},\{2\}} = \frac{12}{3}(1-\frac{1}{2}) = 2.$$

Checking, the numbers not exceeding 12 that are
divisible by 3 but not by 2 are: 3 and 9—Two numbers as
predicted. One more example is

$$12_{\{2\},\phi} = \frac{12}{2} = 6$$

The first subscript indicates 2 from the divisible set. The non-divisible set is the empty set, ϕ. The numbers satisfying are $\{1,2,4,6,8,10,12\}$. Note that some of these are divisible by the prime 3.

Note that $12_{\phi,\{2,3\}}$ are the numbers not divisible by two or three. They are also the numbers that are relatively prime to 12, which is the same as Euler's Totient, function, $\phi(12)$ from Part I of this book. The value of both is 4 and the numbers are $\{1,5,7,11\}$.

Twelve has only two prime factors. With the possibility of the empty set, there are 9 possibilities for our function:

Function and Calculation	Numbers to Check!
$12_{\{2,3\},\phi} = \dfrac{12}{2*3} = 2$	6,12
$12_{\phi,\{2,3\},} = 12(1-\dfrac{1}{2})(1-\dfrac{1}{3})$ $=4$	1,5,7,11
$12_{\{2\},\phi} = \dfrac{12}{2} = 6$	2,4,6,8,10,12
$12_{\{3\},\phi} = \dfrac{12}{3} = 4$	3,6,9,12
$12_{\phi,2} = 12(1-1/2) = 6$	1,3,5,7,9,11
$12_{\phi,3} = 12(1-1/3) = 8$	1,2,4,5,7,8,10,11
$12_{\{2\},\{3\}} = \dfrac{12}{2}(1-1/3) = 4$	2,4,8,10
$12_{\{3\},\{2\}} = \dfrac{12}{3}(1-1/2) = 2$	3,9
$12_{\phi,\phi} = 12$, by convention.	1 through 12

151

Student Problems

1. How many numbers not exceeding a million are

a) divisible by 2 but not by 5?

b) not divisible by 2 or 5?

2. a) In the last table of the text, what is the sum of the counts of all the groups?

b) Make a table for all 9 groups for $n = 15 = 3 \times 5$, just as was done for 12 in the book. Find the sum of all the counts just as in part 2a). The sum is different from 2a), but is there a pattern?

c) Prove in general if $n = ab$, where a and b are primes that the sum of all counts in the 9 groups is 4n. Does it make any difference if there are exponents on a or b ?

3. If n equals the product of three primes, how many groups are there, similar to all the groups in the last

table of the book? What is the sum of the counts of all groups?

<u>Answers</u>

1a) 400,000   b) Also 400,000!

2a) 48 b) 60. Yes, the pattern is that the sum of the counts is four times n.

2c) No

3. 27, $8n$

7. Complex Numbers Applied to Number Theory

Mathematicians created i, the square root of negative one, to solve the simple equation $x^2 + 1 = 0$. There was a world of unforeseen benefits to number theory, engineering and physics. This section shows three small applications to number theory, easily accessible to college algebra students and particularly enjoyable using a calculator that has i. In my first application the students should know about Pythagorean Triples.

Application 1. Finding Pythagorean Triples

The power of complex numbers to find Pythagorean Triples can be easily shown to students by squaring the following 3 complex numbers on a calculator. The calculator must have i = $\sqrt{-1}$.

1. $(2+i)^2 = 3 + 4i$

2. $(3+2i)^2 = 5+12i$

3. $(4+i)^2 = 15+8i$

Now here is the surprise for your students: The a and the b of results a+bi are always the first two of a Pythagorean Triple! In these three cases the triples are (3,4,5), (5,12,13) and (15,8,17). I wouldn't tell the students why for a day or two! The reason is that squaring (a+bi) gives $a^2 - b^2 + 2abi$. But ($a^2 - b^2$, 2ab, $a^2 + b^2$) are a Pythagorean Triple for any integers a and b. In the same spirit of not telling (student discovery), propose the following situations or questions:

1. Find 3 more triples. Include the case b >a. Do you still get the first two of a triple?

2. What conditions on a and b insure that you get a primitive Pythagorean triple? (all three numbers are relatively prime)

3. There are always three different prime numbers that divide into one number of every primitive triple. What are the three prime numbers?

Application 2. The Product of the Sum of Two Squares Equals the Sum of Two Squares in Two Ways

This fact is illustrated with two examples:

Example One.

$$(1^2 + 2^2)(3^2 + 4^2) = 125 = 5^2 + 10^2 = 2^2 + 11^2$$

Example Two.

$$(1^2 + 3^2)(2^2 + 3^2) = 130 = 7^2 + 9^2 = 3^2 + 11^2$$

Complex numbers can give both decompositions into sums of squares! The left side of Example One can be written as:

$$((1+2i)(1-2i)(3+4i)(3-4i).$$

If we take two of these factors that are not conjugates of each other, for example;

$$(1+2i)(3+4i)$$

The result is $-5 +10i$. But 5 and 10 were two numbers, the sum of whose squares gave 125. To find the other decomposition into the sum of two squares just take the conjugate of one of the factors above as in:

$$(1+2i)(3-4i)=11+2i$$

Recall that $125 = 11^2 +2^2$

Application 3. The Triple Product of the Sum of Two Squares is the Sum of Three Squares in Three Ways

Example: I claim that:

$$(1^2+3^2)\ (2^2+3^2)\ (4^2+5^2)= 5330$$

is the sum of three squares in three ways! What are the three ways? Complex numbers to the rescue! Form the product:

$$\left(1+3\ i\right)\left(2+3\ i\right)\left(4+5i\right)=-73+i$$

The result suggests that $5330 = 73^2 + 1^2$. It is! But there are two other decompositions of 5330 into the sum of two squares. Can you find them using complex numbers?

8. How to Win a Snowball Fight

Parametric functions offer a beautiful application to projectile motion and are always covered in my trig classes. If an object is launched from the ground with velocity V and at angle a to the horizontal, then the position at any time t in seconds, is given by the parametric functions:

$$x = V \cos(a)\, t, \quad y = \frac{-gt^2}{2} + V \sin(a)\, t$$

where g is the acceleration of gravity. The origin of our coordinate system is the launch point. We ignore air resistance. We will use these equations for a person throwing a snowball. In letting the launch point be the origin (0,0) of the coordinate system, we are assuming the release height of the snowball is negligible.

It is easy to show (and a good trig application) that the range R is then

$$R = \frac{V^2 sin\ (2a)}{g}$$

So here is the problem that eventually relates to winning a snowball fight. What angle(s) a will give a range equal to 70 percent of the maximum for a fixed velocity V?

The maximum range occurs at 45 degrees and is:

$$\frac{V^2}{g}$$

So, for some angle a:

$$\frac{V^2 sin\ (2a)}{g} = .7\frac{V^2}{g}$$

We have:

$$2a = sin^{-1}(.7)$$

$$2a = 44\ degrees$$

$$a = 22\ degrees$$

The students think we are done, but of course there is another angle that gives 70% of the range. The other angle comes from the idea that the supplement of 44 degrees has the same sine value of .7. So we must also include:

$$2a = 136 \ degrees$$

$$a = 68 \ degrees$$

There are two angles, 22 and 68 degrees that give a range equal to 70 percent of the maximum range. See figure which shows the two trajectories with an initial velocity of 28 m/s. They both give a range of about 56 meters.

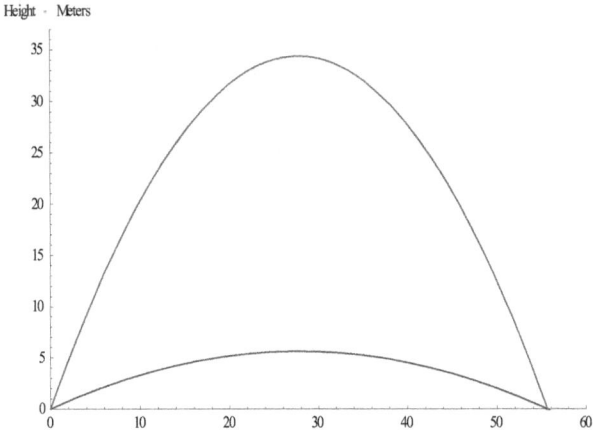

What does this have to do with snowball fights? Well, it's pretty easy to dodge a snowball if you're looking at your opponent. Secretly make two snowballs. If your opponent is at 70 percent of your range throw the first at 68 degrees and while he is looking (and worried about) the high parabolic trajectory, fire the second at 22 degrees. The strategy has a good chance of working. The lower snowball takes less time, so even though it is thrown later it can arrive first at the target!

You might find the times of arrival of both snowballs for V equals 28 m/s and our two angles. You should get that the higher snowball takes about two and a half times as much time—plenty of time to fire the second snowball!

9. How Far Can You See?

Sailors scan the horizon and forest rangers try to see the smallest wisp of smoke in the distance. In many situations we try to maximize the distance to the horizon. This distance is determined by how high you are--the higher the better. But is there a formula for the distance d to the horizon in terms of the observer's height h? This section will find 3 formulas for the distance of increasing accuracy.

The next page has a figure with an observer B at point B and height h above the earth. B can only see as far as horizon point H, due to the curvature of the earth, assumed spherical. Of course, at a greater height you can see further. It's been said that no ship in World War II, accompanied by an observer in a blimp, was ever lost.

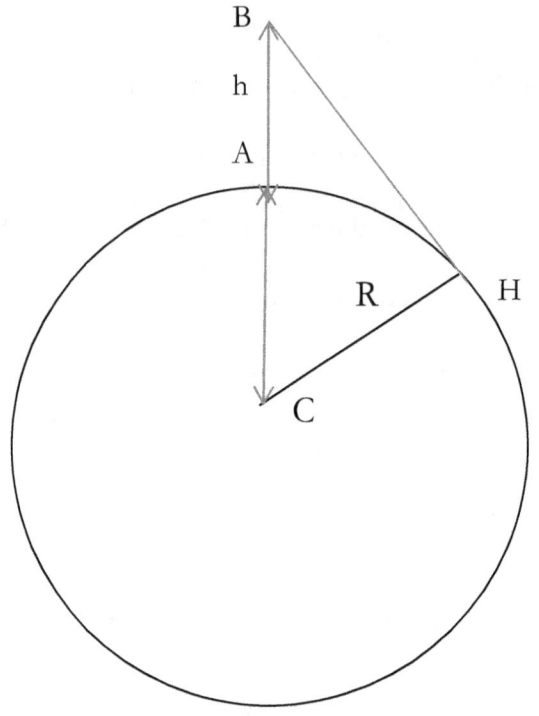

EARTH IN CROSS SECTION—Radius is R, Observer at point B is h above the surface of the Earth

Ideally we want the distance along the earth which is the length of the arc AH. But the straight line distance BH is very close to AH for small h in comparison to R, the radius of the earth. We will find BH as a first approximation to arc AH.

Let C be the center of the circle. BH is easy to find, since triangle CHB is a right triangle (A line tangent to a circle is perpendicular to a radius drawn to the point of tangency). We have, using the Pythagorean Theorem:

$$(BH)^2 + R^2 = (R+h)^2$$

We solve for BH. Many texts drop the h^2 term, since it is usually very small compared to the 2Rh term. Only astronauts have been higher than 1% of R.

$$BH \cong \sqrt{2Rh} \qquad (1)$$

Use formula (1) when h is not a significant fraction of R. Up to a height of 250 kilometers the error in BH, due to dropping h^2, is less than one percent. For higher h, keep the h squared term, giving the formula:

$$BH = \sqrt{2Rh + h^2} \qquad (2)$$

By the way, at 250 km we are in the range of low earth satellites. A later section calculates the time in view of a satellite at this height.

With R for the earth at 6370 km, it's fun to do examples. A kayaker 1 meter high can see 3.6 km. Atop Mt Everest (h = 8.84 km) you can see 335.6 km. to the horizon.

But distance along the earth really means the curved arc AH. We can find that using trigonometry. Let angle BCH be called a. We just need this angle a at the center of the earth. Using the right triangle CHB again:

$$a = sin^{-1}(\frac{BH}{R+h})$$

Then using the arc length formula for angle a in radians:

$$Arc\ AH = R\ sin^{-1}(\frac{BH}{R+h})$$

but BH is $\sqrt{(2RH + h^2)}$

So finally, for all h, the distance you can see along the earth is given by:

$$Arc\ AH = R\ sin^{-1}(\frac{\sqrt{2RH + h^2}}{R+h})\quad (3)$$

There is very little difference between formula (1) and (3) for the kayaker. For the Everest case there is a difference of only .2 km ! We see that the straight line approximation, with the dropping of the h^2, is still very accurate at the height of Everest. Have your students go higher comparing all three formulas (1),(2) and (3)!

The formulas work for any sphere, say on the moon or any planet. We can also look at limiting values in formula (3). As h goes to infinity, the farthest you can see in one direction should be one-fourth of the way around the earth. Finally we can ask students what happens to the distance you can see as the radius of your sphere goes to infinity.

10. How Long is a Satellite in View?

For fifty years the people of earth have rocketed objects into orbit. Some of the thousands of objects circling our globe are visible with binoculars or the naked eye as they traverse the night sky. It's fun and challenging to sit in a lounge chair on a clear night staring, with binoculars handy, toward the right point on the horizon, waiting and waiting. There it is, a manmade moon, streaking across the sky, what satisfaction! It's amazing how quickly it moves, sinking below the opposite horizon, and going under the earth to reappear again in the same place on its next orbit--another quick trip around the world! How long does it take a satellite to streak across our visible portion of the night sky?

The time in view depends of course on the total time for the orbit around the earth. We will find this time first and then use it to find the time in view.

Kepler's law obeyed by all planets is:

"The square of the time to orbit the sun varies directly with the cube of the mean distance from the sun."

This rule is a consequence of the universal law of gravitation and therefore also applies to objects orbiting the earth under our gravitational pull. The translation of Kepler's law into mathematics is:

$$T^2 = kd^3$$

Where T is the time to orbit the earth and d is the average distance of the satellite from the center of the earth. The constant k can be found from one example. What better one then our own, faithful moon? For the moon we have T = 660 hours and d = 390 thousand kilometers. Using hours and thousands of kilometers for our units we get a k of about .007. Thus, for anything orbiting the earth:

$$T^2 = .007d^3$$

$$T = .084d^{1.5} \tag{1}$$

By the way, the mean distance d is the average of the apogee and perigee, both measured from the center of the earth. The radius of the earth is 6.37 thousand kilometers. The early satellites had low heights, say from 200 to 300 kilometers above the surface of the earth. At an h of 250 km., the mean distance from the center is 6.62 thousand kilometers (counting the radius of the earth) and the time to orbit by equation (1) is 1.43 hours. Low satellites have a quick trip around the world, but we'll see that the time in view for our nighttime loungers is much shorter!

For the time in view let's consider the simplest case--the satellite passes directly overhead and the orbit is close to a circle (low eccentricity). The later condition ensures that the satellite does not change speed significantly. In the figure below, the radius of the earth is R and the satellite is at an

average height h above the earth.

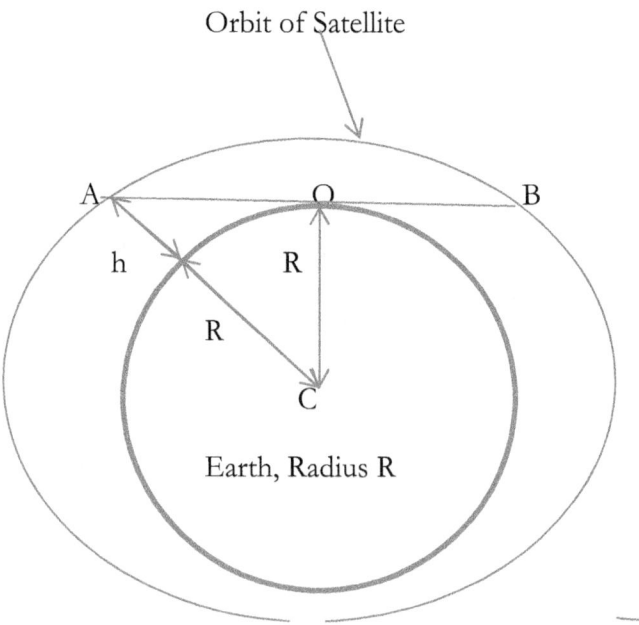

Orbit of Satellite

SATELLITE IN CLOCKWISE ORBIT.

OBSERVER AT POINT O FIRST SEES

SATELLITE AT ORBIT POINT A, THEN

LOSSES SATELLITE AT POINT B.

The lines from the observer at O to the horizon at point A or the opposite horizon point B give the very lowest points that we can hope to see the satellite. Our little artificial moon is only in sight from point A to point B in its orbit. If we can calculate what fraction the Arc AB is of the whole orbit then we can find the time in view. Let C be at the center of the earth. Triangle AOC is a right triangle. Take h as the distance from point A to the earth. The angle ACO is $\cos^{-1}(R/(R+h))$. One slight problem is that h is changing, but a variation of 100 kilometers is very little when measured from the center of the earth. The eccentricity of the orbit is small and the angle will be pretty accurate if we use an average height of 250 kilometers. Angle ACO is

$\cos^{-1}(6370/(6370+250))$ equals about 16 degrees. We must double this to get the total angle of viewing, 32 degrees. Finally, use a proportion to find the viewing time.

Viewing Time:

$$= \frac{\text{Total Angle in View}}{360} \times \text{Orbital Time}$$

$$= \frac{32}{360} \times 1.43 \ hrs = .13 \text{ hours}$$

That is less than eight minutes for this low orbit satellite! The actual time in view will always be a little less since horizons are usually obscured by trees, hills or buildings. So if your satellite viewing party is slow getting outside you will have to wait another hour and a half to try again!

Can your students generalize this theory to find T, the time in view for a height h? The answer is:

$$T = \frac{2cos^{-1}\left(\frac{R}{R+h}\right)}{360}(.084) \ (R+h)^{1.5}$$

where T is in hours, degrees are used in the cosine inverse, R and h are in thousands of kilometers. This uses equation (1) and assumes the mean distance d of Kepler's law is R +h.

One last point is that this calculation ignores the rotation of the observer on the earth. In the most general case there are many other factors. A non-circular orbit and changing height causes the satellite to change speed—We'd have to know the height and changing speed at our location which could change with each orbit-- quite complicated!

As I write this (Sept, 2011), the Upper Atmosphere Research Satellite (UARS) is due to crash into earth. NASA states a most likely estimate of 48 hours until the crash. But they are putting a possible error of ±14 hours in this estimate. The best estimate for the location of the crash is a circle of radius no less than 1000 miles! In this case, of course, Kepler's law does not apply. Also the tumbling of many differently shaped pieces of debris interacting with air resistance make accurate predictions impossible.

11. The Many Uses of Infinite Geometric Series

We all know that for $|r| < 1$:

$$1 + r + r^2 + r^3 + \dots = \frac{1}{1-r}$$

I have three favorite applications of this remarkable formula.

Application 1. Bouncing Ball, Total Distance Travelled and Total Time in Infinitely Many Bounces

Assume a ball is dropped from a height h and rebounds to the fraction r of the falling distance. Assume that this rebounding ratio r continues for infinitely many bounces. Then the total distance D that the ball travels in the infinite bounces is:

$$D = h + 2hr + 2hr^2 + 2hr^3 + \dots$$
$$= h + \frac{2hr}{1-r}$$

Do a numerical example for the students.

An object dropped from a height s will take $\sqrt{\dfrac{2s}{g}}$ to reach the ground, where g is the acceleration due to gravity. It also takes this long to rise to height s from the ground.

Suppose we drop the ball again from height h but now are interested in the total time that the ball bounces in infinitely many bounces—it won't be infinite! As before, we assume the ball always rebounds the fraction r. So the total time T that an object will take to do infinitely many bounces is:

$$T = \sqrt{\frac{2h}{g}} + 2\sqrt{\frac{2hr}{g}} + 2\sqrt{\frac{2hr^2}{g}} + 2\sqrt{\frac{2hr^3}{g}} + \ldots$$

$$= \sqrt{\frac{2h}{g}} + 2\sqrt{\frac{2hr}{g}}(1 + r^{1/2} + (r^{1/2})^2 + (r^{1/2})^3 + \ldots$$

$$= \sqrt{\frac{2h}{g}} + 2\sqrt{\frac{2hr}{g}}\left(\frac{1}{1-r^{1/2}}\right)$$

For h = 2 meters, g = 9.8 meters/s^2 and r = .6, the total

distance the ball travels is 8 meters and the total time is about

5.03 seconds. The average velocity is 1.59 m/s, which is

coincidentally close to $\pi/2$!

Application 2. The Total Economic Impact of a New Manufacturing Company on the Local Economy.

A new company comes to town with an annual

payroll of P. Suppose the company is unique to the city so

that it does not compete with other businesses. Assume that

all people in the city spend the fraction r of their income in

the city. So rP is new income for local people who in turn

spend r of it (r^2P) locally. This continues forever and is

called the multiplier effect in economics. Let I be the total

impact on the local economy. Then:

$$I = \text{Pr} + \text{Pr}^2 + \text{Pr}^3 + \ldots$$
$$I = \frac{\text{Pr}}{1-r}$$

For an annual payroll P of 10 million dollars and an r

of .65, the impact to the local economy is 18.6 million dollars,

almost twice the payroll. If a 10 million annual payroll

company shuts down then, unfortunately there is a loss to the

local economy of 18.6 million dollars.

Application 3.Infinite Annuities

In finance theory, the present value P of one payment in the

future is given by

$$P = \frac{F}{(1+r)^n}$$

where F is the future payment, r is the interest rate per period

and n is the number of periods compounding into the future.

Now consider a loan. There will be n payments of p

(lower case) and each payment has a different present value

because the creditor must wait a different length of time for

each payment. The present value P of the entire loan will then

be the sum of the present values of the n payments. Assume

the first payment is due after one period. The present value P

of all n payments will be the sum of the n terms as follows:

$$P = \frac{p}{(1+r)} + \frac{p}{(1+r)^2} + \frac{p}{(1+r)^2} + \ldots + \frac{p}{(1+r)^n}$$

$$P = \frac{p}{1+r}(1 + \frac{1}{(1+r)} + \frac{1}{(1+r)^2} + \ldots + \frac{1}{(1+r)^{n-1}})$$

$$P = \frac{p}{(1+r)}(\frac{1 - \frac{1}{(1+r)^n}}{1 - \frac{1}{1+r}})$$

*Here we used the formula for a finite geometric
series.*

This simplifies as follows.

$$P = \frac{p}{r}[1 - \frac{1}{(1+r)^n}] = \frac{p}{r}[1 - (1+r)^{-n}] \quad (PV)$$

Call this formula (PV), for present value.

Formula (PV) gives the present value of n payments

stretching into the future.

We can show our classes a dramatic change in the present value as the length of the loan increases. I will even extend the loan to infinitely many payments!

Suppose a particular car is offered at payments of $300/month for 5 years (n=60). Let r be .5% per month. The 'amount' of car you can buy is the present value P, which using our (PV) formula is $15,518. But what you if want a more expensive car? You may be able to get longer term financing-- no problem in a in a math class! An n equal to 120 (10 years), p = $300 and r = .005, as before gives a present value P of $27,022. You could buy a $27,022 car but would be stuck with 10 years of monthly payments! At about $27000, it should be a lot nicer car, but how nice would it be by the time you paid it off?

Discounting that practical consideration, let's extend our $300 payments arbitrarily far into the future. We get the table below. Note the third column. As time increases the sum of payments—the total you pay increases linearly. The third column also has the ratio of the sum of payments to the

present value. This ratio is increasing much faster than the

total sum of payments. As the number of payments increases,

you are paying much more over time, compared to the

present value of the car. Remember that the present value is

the value of the promise to pay, the loan contract. This

monetary amount is also assumed equal to the car's value

today.

<u>Present Values of $300/Month at .5% Compounded</u>

<u>Each Month For Increasing Time Periods</u>

Number of $300 Monthly Payments	P=Present Value (Car's Value Now)	Total Sum of Payments, and (Ratio to P)
60 (5 years)	$15,517	$18,000, (1.16)
120 (10 years)	$27,022	$36,000, (1.33)
240 (20 years)	$41,874	$72,000, (1.72)
480 (40 years)	$54,524	$144,000, (2.64)
960 (80 years)	$59,500	$288,000, (4.84)
1920 (160 years)	$59,996.59	$576,000, (19.2)
Infinitely Many!	$60,000	Infinite!, (Infinite)

Consider the last row, infinitely payments. At n equal to infinity our (PV) formula becomes:

$$P = p(\frac{1-0}{r}) = \frac{p}{r} \quad ,$$

The present value with infinitely payments is simply p/r.

In our case, this is:

$$P = \frac{\$300}{.005} = \$60,000$$

Notice how close this present value (The value of a contract to pay $300/month forever!) is to the contract to $300 for 'only' 160 years. The 'difference' between 160 years of payments and an infinity of payments is only $3.41!

Also note that the interest before the first monthly payment is made on our $60,000 car is (.005)60000 = $300. This equals the whole monthly payment. Thus the $300 payments are all interest and, by all rights, the payments must go on forever!

12. Easy Ways to Become a Half Millionaire (Discrete Deposits vs. a Continuous Stream of Deposits)

Depositing money periodically into a savings account is the safest way to accumulate money. If done over a long time it is also the easiest way to accumulate a lot. Suppose we deposit money every year for 40 years at 6% compounded yearly. There could be an employer program where the first deposit is made at the end of one year and the 40th deposit is made on the last day of work, after 40 years. How much needs to be deposited to have $500,000 at the end?

But wait, suppose we deposit money at the end of the first month and every month thereafter for 40 years? The money is now hitting the bank sooner. The total amount to reach $500,000 should be less than waiting for the end of each year. Throughout the 40 years there will be more totally deposited at any time except the end of a year.

The total needed will be even less if we deposit money every week or every day! The theme of this section is finding the minimum total to deposit as the frequency of depositing increases without limit. We will see that calculus provides the theoretical minimum. Hopefully, this will make a moral for our young students. Begin the deposits while you are young!

Here is the derivation of the future value F of a series of n payments of size p made every period. The frequency of compounding will always match the frequency of deposits. The interest rate r will always be .06 (6%) divided by the number of times compounded per year. The first deposit is made after the first period and the last deposit is made at the very end of 40 years and earns no interest. Then:

$$F = p + (1+r)p + p(1+r)^2 + \ldots + p(1+r)^{n-1}$$
$$F = p[1 + (1+r) + (1+r)^2 + \ldots + (1+r)^{n-1}]$$
$$F = p\frac{(1+r)^n - 1}{r}$$

In our case F =$500,000 and p is desired. So

$$p = \frac{rF}{(1+r)^n - 1} \qquad \text{(Discrete Payments Formula)}$$

For yearly deposits, n is 40, and r = .06. The formula

gives p as $3230.77. These are the yearly deposits. The total

deposited in 40 years would be $129,230.80.

How much less if we deposit money every month?

We change n to 40 times 12 equals 480 and r, the interest rate

per month is now .06/12 equals .005. The result is a p of

$251.07 per month. The total paid now to reach 500 grand is

$120,512.74, about $9000 less than before.

Rather than continuing with arbitrarily more frequent

deposits, let's use calculus to find the ultimate minimum

amount deposited to reach our goal. An integral requires a

function, a continuous stream of money (money/time). Also the money has to be compounded continuously to match our rate of depositing, as in the discrete case.

We use a continuous rate of interest r. Assume p is the constant rate of continuously deposited money per year. So, p is in the units: dollars/year. An infinitesimal amount of money is pdt. If t is the time elapsed and T is the total time that money is deposited, then (T-t) is the time that the pdt earns continuous interest . So at the end of time for the annuity one small piece of money is worth:

$$pdt \times e^{r(T-t)}.$$

Each small piece of money is in the bank for a different amount of time. We must sum all these small amounts of money from t =0 to t equals T. Thus we have use the integral:

$$F = \int_0^T p e^{r(T-t)} dt$$

Solving for p:

$$p = \frac{F}{\int_0^T e^{r(T-t)} dt} = \frac{F}{\int_0^T e^{rT} e^{-rt} dt}$$

$$p = F / (e^{rT} \left. \frac{-1}{r} e^{-rt} \right|_0^T)$$

$$p = F / (e^{rT} \left. \frac{-1}{r} e^{-rt} \right|_0^T)$$

$$p = \frac{Fr}{e^{rT}(1 - e^{-rT})}$$

$$p = \frac{Fr}{e^{rT} - 1}, \qquad \text{(Payments Formula--}$$

Continuous Deposits)

.

The ultimate lowest that needs to be deposited is then

2993.06 dollars/year times 40 years or $119,722.40.

We see that this ultimate is lower than the monthly

rate by only $790, pretty disappointing over 40 years.

Continuous depositing is of course not practical—we would have to feed money in continually 24--7 for 40 years! The theory of continuous compounding is practical as a lower limit on what needs to be deposited as the frequency goes to infinitely often.

I summarize everything and add a few other frequencies in the table below.

Payments and Total Paid to Reach $500,000 in 40 Years. Interest is 6%

Freq. of Deposits	p=Deposit	Total Deposited
Yearly	$3231	$129,240
Semi-Annually	$1556	$124,480
Quarterly	$763	$122,080
Monthly	$251.07	$120,513
Daily	$8.1963	$119,748
Continuous Stream	$2993.063/Year or $8.1946/day	$119,723

For the daily figures a year of 365.25 days was used. The total paid shows big changes at first—almost $5000 less in going from yearly to semiannually. But from quarterly to daily the total paid only decreases by $2,278. Daily is 90 times as often as quarterly! The difference between daily and continuously is absolutely trivial at 6% interest, $25 over 40 years—less than one-fifth of a cent per day!

Calculus and continuous compounding provide very good limits on the lowest discretely deposited amounts which are needed to reach a financial goal.

13. Areas of Parabolic Segments

There are at least three ways to find the area of parabolic segments. I think we should show our calculus students these ways.

Consider the area bounded by $y = 16 - x^2$, and the $x -$ axis:

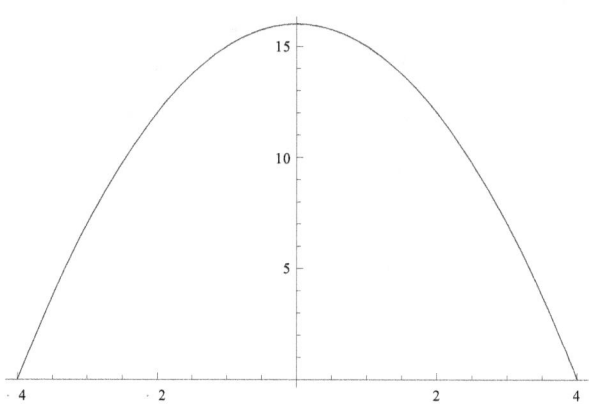

The three methods follow.

Method 1. Integrate with respect to x:

$$\int_{-4}^{4} 16 - x^2 dx = 2\int_{0}^{4} 16 - x^2 dx = 85\frac{1}{3}$$

Method 2. Integrate with respect to y:

$$\int_{0}^{16} 2\sqrt{16 - y}\, dy = -\frac{4}{3}(16 - y)^{\frac{3}{2}}\, |_{0}^{16} = 85\frac{1}{3}$$

Method 3. Archemedes Way

Let's not forget Archimedes who found and proved that the area of a parabolic segment was equal to 4/3 rds the area of the inscribed triangle. So:

$$A = \frac{4}{3}A_t$$

$$= \frac{4}{3}*(\frac{1}{2}*8*16)$$

$$= 85\frac{1}{3}$$

It is very important to make the student draw separate pictures for the two different integration methods. A representative vertical rectangle in method 1 and a horizontal one for method 2 should be drawn. In most cases it is the algebra, for example finding the width function in method 2, that students need the most practice with. After seeing the three ways to find the area of a parabolic segment, how could anyone not be impressed with calculus!—and, of course with Archimedes.

14. *Minimizing Perimeter*

Suppose a circle and a square have equal perimeter. The circle has greater area, of course. My College Algebra students have a hard time proving this. They are not used to manipulating two or more formulas in one problem. If P is the common perimeter, we get the areas:

Square: $\dfrac{P^2}{16}$

Circle: $\dfrac{1}{4\pi}P^2$

Can students state the relative advantage of the circle? It's about 21% based on the circle's area .

I think this general result, comparing all squares to all circles is a great use of algebra. Problems like this are not in our texts. We spend so much time solving specific quadratic equations or teaching so many different topics from absolute value to logarithms. This interplay of geometry and algebra is so fruitful that I would like to continue with this theme of areas with circular boundaries versus areas with straight edges. In the variations that follow we try to maximize the area for a fixed amount of fencing P.

Problem 1. One side of a field does not have to be fenced. If P fencing is available find the dimensions of the rectangle with greatest area and compare to a semicircle with the diameter along the side that does not have to be fenced.

<u>Answer and Comment</u>:

The largest rectangle is: $\dfrac{P}{4}$ by $\dfrac{P}{2}$ with area $\dfrac{P^2}{8}$

A semicircle with radius of $\dfrac{P}{\pi}$ has an area of $\dfrac{P^2}{2\pi}$

The circular area wins again and by the same relative amount as before, 21%!

In standard text book problems, the college algebra student is expected to maximize a quadratic function by finding the vertex. Students have trouble using the constant P rather than a specific amount. They have even more trouble comparing the best rectangle to a semicircle. If we don't use variables such as P and if we don't ask our students to compare rectangles to circles they miss the chance to practice algebra, and see general results. Here is another example comparing a region with straight sides to a one with a circular boundary.

Problem 2. A region with two equal, separated pens is be formed. Maximize the total area which has a divider separating the two pens. The divider is to be the common border of the pens and is part of the total fencing available, P. Try a rectangle with total fencing P forming the four sides and the divider. The curved area competitor will be a circle with the same total fencing P including a divider separating

the pens. The divider will be along the diameter of the circle.
How do the total areas compare?

<u>Answer and Comment:</u>

The greatest rectangle: has dimensions:

$$\frac{2}{9}P \text{ by } \frac{P}{6} \text{ with area } \frac{P^2}{27}$$

Circle: Radius is $\dfrac{P}{2(\pi+1)}$ with area $\dfrac{\pi}{4(\pi+1)^2}P^2$

The circle wins again but the relative advantage is a little
lower, 19%. Here is one more example requiring calculus.

Problem 3. Three equal area pens are required with
one straight side along an existing wall that does not have to
be fenced. Use a semicircle with a diameter as the non-fenced
side. Use straight line dividers. Let P be total fencing. How
should the three regions be formed so as to maximize the
total area? Compare semicircles divided vertically and

horizontally to a rectangle with two dividers perpendicular to the unfenced side.

Answer and Comment

With P as the total fencing, three cases (not exhaustive) are shown below. The total areas are given in terms of P. Calculus was used to ensure that the three regions have equal area. Remember that fencing had to be used for the curved upper boundary and all the dividers, but not for the diameter.

Case One: Two Vertical Dividers:

$Area \cong .061P^2$ (Maximum of the Three Cases)

<u>Case 2: Three dividers. Leftmost and rightmost areas make</u>

<u>up one of the three regions.</u> *Area* $\cong$ $.047P^2$

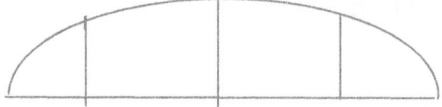

<u>Case 3: Two Horizontal Dividers—The least area case</u>

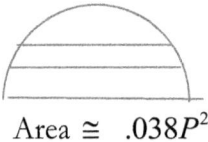

Area $\cong$ $.038P^2$

Finally, the best rectangle with one side unfenced, two

dividers, and three equal regions has area $\dfrac{P^2}{24} \cong .042P^2$.

Thus, two of the circular cases beat the best rectangle.

Is it possible for a rectangular area to beat the best

circular area? I can't find a case!

15. Differentials and Relative Change

Some calculus texts do not have differentials and almost all do not use them with relative change. A simple example with a simple geometric justification follows.

Consider a square with side s and area A. Using the formula for the area, let us find:

a) dA and b) The relative change dA/A in terms of s.

<u>Solutions</u>:

a) $A = s^2$

$dA = 2s\ ds$

b) We form the ratio:

$$\frac{dA}{A} = \frac{2sds}{s^2} = 2\frac{ds}{s}$$

We interpret this equation for our students, " dA/A and ds/s are the relative changes in the area and in the side respectively." We picture a square growing or shrinking by ds and the area in turn changing. We can draw a square growing for our students. The equation suggests that for small changes in s the relative change in the area is twice the relative change in the side. We also do numerical examples showing close and for larger ds/s not so close agreement to the exact relative change in the area.

We can also think of the ds as an error in measuring the side. Errors, of course, always occur in measuring, since nothing continuous can be measured exactly. Thus dA/A is the relative error in the area when the side is measured with relative error ds/s. Differentials have wide applicability— squares, circles, spheres—all growing, or shrinking or measured with small errors. The gap in calculus teaching is that our books do not force students to form the relative change ratios. Failing to do this, they do not see the

relationship between the relative changes.

Relative errors using differentials are, of course, not

restricted to geometric figures. Consider how g, the

acceleration of gravity varies with the changing distance r,

from the center of the earth. The initial relation is:

$$g = kr^{-2}, \text{ where k is a constant.}$$

$$\frac{dg}{g} = -2\frac{dr}{r}$$

Thus, a 5% increase in your distance from the center

of the earth means about a 10% decrease in g. The actual

change (from a lot more calculation) is a 9.3% decrease. The

approximation using differentials is much easier and faster to

calculate, is fairly accurate over ranges from zero through

about 6% of r, and applies to any planet.

16. Discrete Average versus Continuous Average

The average value of a function over an interval is defined in calculus. Shouldn't we compare this average to the students' previous understanding of average for a discrete set of numbers? We usually don't have the time to fully explore this link. It takes time to do and to make examples that come out nice. Here is one example.

Suppose we want the average value of $f(x) = x^2 + 5$ over the interval 0 to 4. The continuous (calculus) average of a function $f(x)$ over the interval (a,b) is defined as:

$$\frac{\int_a^b f(x)dx}{b-a}$$

Our case becomes:

$$= \frac{\displaystyle\int_0^4 x^2 + 5 \, dx}{4 - 0}$$

The continuous average works out to 10 1/3. The following table gives the averages for different numbers of discrete values of (x^2+5) uniformly distributed from x = 0 to 4. The last row shows the continuous-calculus based average for comparison

n (x Values)	Average$= \dfrac{\sum_i x_i^2 + 5}{n}$
2 (at x = 2 and 4)	15
5 (x= 0,1,2,3 and 4)	11
9 (x= 0,.5,1,1.5,...4)	10 2/3
21 (x=0,.2,.4,.6,...4)	10.46666...
Infinitely many n	10.3333...--Thanks to Calculus

It doesn't take a high value of n to get pretty close to the calculus average. I just love putting, "Infinitely many n", in the last row. A center of mass example follows, again comparing a continuous and a discrete situation.

Center of Mass Example—Discrete Masses Versus Continuous Mass

Consider a thin rod with linear density kx where k is a positive constant with the units kg/m and x is the distance from the origin in meters. Then the center of mass, c for a rod 5 meters long is found as follows:

$$c = \frac{\int_0^5 x \bullet (kx)dx}{\int_0^5 (kx)dx}$$

$$= 3\frac{1}{3} \text{ meters}$$

This center of mass or balance point is closer to the right end as expected for a rod whose density is increasing with x. Our result is independent of the constant k.

Now, can we give our students a discrete example that looks like it is converging to this result? Yes, suppose 5 masses, denoted as (location,mass):

(1m,1kg.), (2m,2kg.), (3m,3kg.), (4m,4kg.), and (5m,5kg.).

Note that the discrete masses are increasing linearly to the right, similar to the continuous case.

Then using the center of mass formula for a discrete system of masses we have:

$$c = \frac{\sum_{i=1}^{5} m_i x_i}{\sum_{i=1}^{5} m_i} = 3\frac{2}{3} \; meters$$

Already we are pretty close to the continuous case. I can't resist getting closer. Suppose now 10 mass at locations and of mass sizes as follows:

(.5m,.5kg),(1m,1kg),(1.5m,1.5kg),....(5m,5kg). Then

$$c = \frac{\sum_{i=1}^{10} m_i x_i}{\sum_{i=1}^{10} m_i} = 3\frac{1}{2}$$

So without too much trouble, we are showing that finding the center mass of discrete masses is approaching quite closely the center of mass of a continuous rod. My aim is make calculus students more accepting of the calculus answer.

17. Variations on the Identity Matrix

It is instructive to make slight changes in the 2 x 2 Identity matrix and ask what affect it has when it is multiplied with an arbitrary matrix. For example, ask the students to multiply the matrices below and more importantly to write the effect on the arbitrary matrix to the right.

$$\begin{bmatrix} 1 & 1 \\ 0 & 0 \end{bmatrix}\begin{bmatrix} a & b \\ c & d \end{bmatrix}$$

The result is:

$$\begin{bmatrix} a+c & b+d \\ 0 & 0 \end{bmatrix}$$

We expect the students to write, "The effect on the arbitrary matrix is that the first row becomes the sum of the first two rows and the second row is zeroed." At least something like this is expected. What if the altered Identity matrix above is right multiplied with an arbitrary matrix?

If the entries of an altered 2 by 2 Identity matrix are either 0 or 1, then there are $2^4 = 16$ possible matrices, including the identity matrix itself. Have the students explore the effects of these matrices when multiplied (on left or right) with an arbitrary matrix. After a while, the students should try predicting effects and verifying them.

One can also consider non-square identity type matrices. For example, what is the effect when the matrix

$$\begin{bmatrix} 1 & 0 \\ 0 & 0 \\ 0 & 1 \end{bmatrix}$$

is left multiplied with an arbitrary (2×2) matrix? The above matrix can be left multiplied with an arbitrary $(2 \times n)$ matrix. Letting the entries be larger than 1 provides even more variations.

Multiplying 'Altered Identity Matrices' with points in the plane and with a set of points defining a polygon provides a visual context for the effect of these matrices. I think these are very good exercises for understanding what matrix multiplication means.

www.ingramcontent.com/pod-product-compliance
Lightning Source LLC
Chambersburg PA
CBHW071714170526
45165CB00005B/2008